COMBUSTION ENGINES—REDUCTION OF FRICTION AND WEAR

IMechE CONFERENCE PUBLICATIONS 1985–3

Sponsored by
The Power Industries Division of
The Institution of Mechanical Engineers and
The Japanese Society of Mechanical Engineers

18–19 March 1985
1 Birdcage Walk, Westminster, London SW1

Published for
The Institution of Mechanical Engineers
by Mechanical Engineering Publications Limited
LONDON

First published 1985

British Library Cataloguing in Publication Data

Combustion engines: reduction of friction and wear: papers read at the conference held at the Institution of Mechanical Engineers, London on 18–19 March 1985.
1. Combustion engineering
I. Institution of Mechanical Engineers, *Power Industries Division* II. Nihon kikaigakkai
621.402′3 TJ254.5

ISBN 0 85298 559 2

Printed by Waveney Print Services Ltd, Beccles, Suffolk

CONTENTS

C67/85

Friction in internal combustion engine bearings

F A MARTIN, CEng, FIMechE
The Glacier Metal Company Limited, Ealing Road, Alperton, Wembley, Middlesex

SYNOPSIS

Bearing friction forms a significant part of the total mechanical losses within an engine. Reduction in these losses will help with energy conservation. Various techniques, used internationally, for predicting bearing friction are reviewed. The effect of bearing size, bearing shape, oil feed grooving, viscosity and environmental effects are discussed. Many of these factors are incorporated in power loss prediction charts and trend analysers. Examples of friction reduction in redesigned engines are given.

1. INTRODUCTION

The need for energy conservation has meant that a deeper understanding is required of the mechanisms of friction losses within all components of an internal combustion engine. In the automotive field it has been estimated that a 10% reduction in the direct mechanical losses within an engine gives perhaps 1 to 3% improvement in fuel consumption. Such savings are very significant when viewed on a worldwide basis, in both economic and conservation terms. For these reasons many manufacturers, particularly in Japan and USA (1, 2, 3, 4) have gained improvements in fuel economy by reducing the mechanical friction losses within the engine by careful re-design. Bearings are one component which contributes to the overall friction losses, and this paper reviews the current knowledge on bearing design as it relates to friction losses, and attempts to correlate the work of many researchers.

The calculation of friction losses within a bearing oil film is an integral part of the design process for that bearing. These losses, appearing as heat, raise the temperature of the lubricant within the clearance space, and lower its operating viscosity. This change in viscosity heavily influences the oil film thicknesses within the bearing throughout the entire journal orbit, which in turn feed back into modified power losses. Therefore an accurate assessment of bearing friction is a necessity in its own right when carrying out any detailed design work on engine bearings. Techniques for predicting such orbits were described by the author in a basic review (5) which has recently been updated (6). In order to be able to compare different prediction methods a particularly well documented bearing has been taken as a study case in these and other papers. This same bearing (the big end of a Ruston and Hornsby VEB MK III 600hp 600 rev/min diesel engine) has been used in this present review paper to compare friction calculation methods. In particular, studies have been carried out by the Glacier Metal Company and GEC in England, and General Motor Research Laboratories in the USA. Also fundamental data from Delft and Twente Universities in the Netherlands and Cornell University in the USA have been used.

In order to give some help with the initial design of bearing systems, an attempt has been made to give guidance on the effect of various bearing and environmental factors on friction losses. In this work the basic bearing dimensions of diameter, length and clearance have been related to the choice of lubricant viscosity within the bearing, and the relative values of the rotating and reciprocating masses within the engine. Prediction charts are presented which are based on many diverse theoretical studies, and comparisons are made with experiments on motored engines.

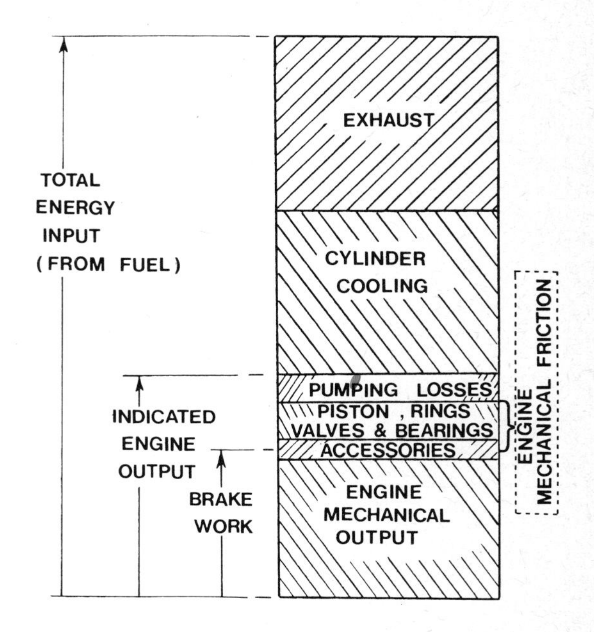

Fig 1 Typical energy distribution in an automotive engine
(Petrol engine — part open throttle)

2 NOTATION

General

b	=	bearing length	(m)
c or c_r	=	radial clearance	(m)
C_d	=	diametral clearance	(m)
d	=	bearing diameter	(m)
e	=	eccentricity	(m)
F	=	bearing load	(N)
F_F	=	friction force	(N)
h	=	film thickness	(m)
H	=	Power loss	(W)
N	=	Rotational speed	(rev/s)
P	=	Specific load	(N/m^2)
R	=	Crank throw	(m)
η	=	Operating viscosity	(Ns/m^2)
θ	=	angle from h_{max}	(deg)

Relating to Power Loss Equation – (Tables 1, 2, Fig 6)

H_{ROT}	=	Power loss due to rotation	(W)
H_{TRANS}	=	Power loss due to translation	(W)
H_{TOTAL}	=	$H_{ROT} + H_{TRANS}$	(W)
F_j	=	Force applied to oil film by the journal	(N)
τ_j	=	Torque exerted on the oil film by the journal	(Nm)
τ_b	=	Torque exerted on the oil film by the bearing	(Nm)
ω_j	=	Angular velocity of journal	(rad/s)
ω_b	=	Angular velocity of bearing	(rad/s)
ϕ	=	Angle measured from F_j to h_{min} position	(deg)
β	=	Angle measured from F_j to $\underline{V}$	(deg)
V	=	Journal centre velocity (ref x,y frame)	(m/s)

Relating to Inertia Study (Figs 8 and 9)

A, B and C – load diagram shape factors

F_c	=	Rotating inertia force	(N)
P_c	=	Specific load $F_c/(bd)$	(N/m^2)
M_c	=	Rotating mass	(kg)
M_I	=	Reciprocating mass	(kg)

Dimensionless Terms (in consistent units)

Coefficient of friction $\quad f = \frac{F_F}{F}$ or $\frac{F_F}{F_c}$

Dimensionless power loss $\quad H' = \frac{H\,C_d}{\eta\, b d^3 N^2}$

Load Number $\quad W'_c = \frac{P_c}{\eta N}\left(\frac{C_d}{d}\right)^2$

$$= \frac{2\pi^2 M_c N R}{b r \eta}\left(\frac{c}{r}\right)^2$$

Eccentricity Ratio $\quad \epsilon = e/c_r$

3. BEARING FRICTION AS PART OF THE ENERGY DISTRIBUTION

The energy distribution and friction in various engines have been the subject of detailed study by many researchers (1) (2) (3) (4) (7) (8). Information has been presented on both petrol and diesel engines, on low and high load conditions, and on part open and full open throttle conditions. More general reviews on the subject are given by Rosenberg (9) and Parker and Adams (10).

The mechanical friction losses in an engine are developed in the piston (skirt) and rings, the valve system, the bearing system, and those accessories directly associated with the engine. A typical energy distribution (see also Pinkus and Wilcock (11) Lang (12)) is shown in Fig 1 for a passenger car engine under part throttle operating conditions. This figure shows schematically that about 30% of the energy available within the fuel goes out to the exhaust, 30% is involved in cylinder cooling, with only 40% providing the indicated output. However, only 25% of the total energy is ultimately available as brake work (to transmission and road wheels) with mechanical friction and pumping losses accounting for the remainder. Summarizing, the engine mechanical losses account for –

8 to 10% of total energy input (total energy basis)
or 20 to 25% of indicated output (indicated output basis).

These figures cannot be totally general, and they will vary with operating conditions, type of engine etc. However, they do serve to show the scale of the friction loss within the engine. More detailed effects are shown by the experimental work at the Ford Motor Co (4) which illustrates the effect of full and part throttle conditions on mechanical friction. This is reproduced in Fig 2, and shows a greater proportion of mechanical losses at part throttle conditions; it indicates how the various engine components contribute to the friction losses, although pistons and bearings are considered together.

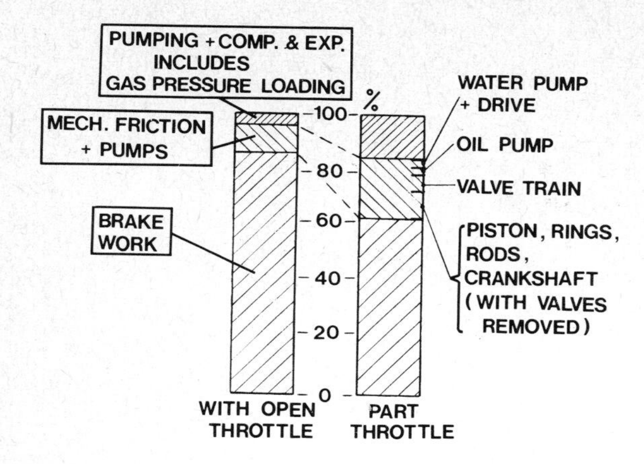

Fig 2 Effect of full and part throttle on engine friction Ford Motor Company — Kovach et al (4)

Dr Lang of Daimler Benz, Germany, presented an overview of 'Friction losses in combustion engines' at the 1981 'Limits of Lubrication' Conference, and this talk has since been published (12). Lang states that the mechanical losses are approximately 25% of the energy in the cylinder, giving details in an excellent pictorial presentation, Fig 3. This is based on average values for engines with different numbers of cylinders, auxiliaries, types of combustion, speed and load. Some of these mechanical losses are attributed to auxiliaries such as water and oil pumps, cooling fan, generator, compressor and hydraulic pumps etc. According to Lang, friction losses which could be improved or optimised tribologically amount to about 72.5% of the mechanical losses, consisting of:-

valve train	6%
connecting rod bearings	10%
main bearings	12.5%
piston rings	19%
pistons	25%

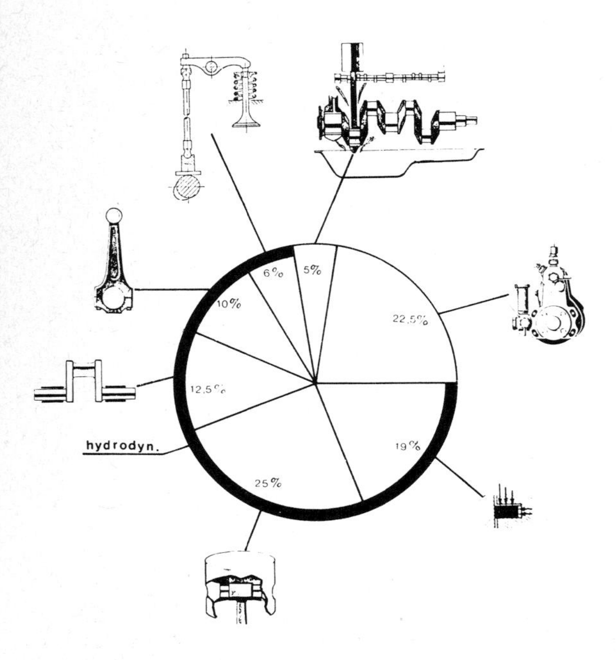

Fig 3 Mechanical losses in combustion engines, Lang (12)

Typical distributions of friction loss, associated with various engine components, recently formulated in the USA (4) and Japan (3) are shown in Fig 4 (a) and (b). These trends, showing friction mean effective pressure (proportional to friction torque) against engine speed, are from motored engines. It is difficult to see exactly the friction loss contribution of the bearings alone, since in the Ford experimental results, Fig 4 (a), seals are included with the crankshaft, and the piston assemblies are included with the connecting rod bearings. However, they do give a useful guide. The experiments from Fuji Heavy Industries Ltd, Fig 4(b), are more enlightening, since the losses within the piston system (generally forming a major part of the mechanical losses (12)), are considered separately.

Many other studies on motored friction (2) (7) (9) (10) (11) (13) are reported in the literature; in particular the work of Bishop (8) has formed a basic reference work over the past two decades. Taking a logical grouping of variables and using regression analysis techniques he gives a broad analytical guideline for a 'rough' but reasonable approximation of the various friction components, including the crankcase mechanical friction.

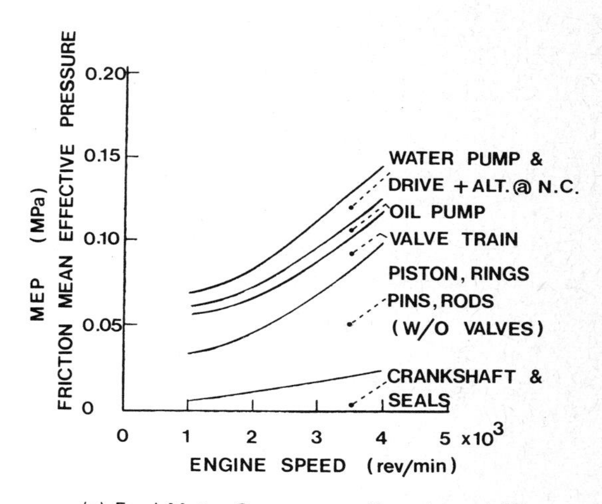

(a) Ford Motor Company — Kovach et al (4)

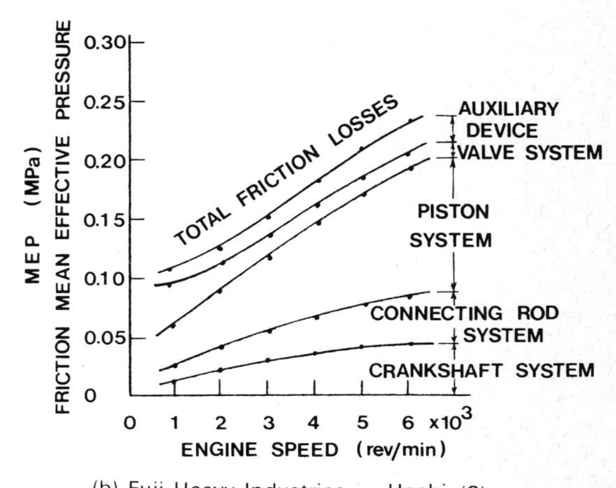

(b) Fuji Heavy Industries — Hoshi (3)

Fig 4 Examples of engine component friction from motored engines

4. PREDICTION OF BEARING FRICTION AND POWER LOSS

4.1 Some basic concepts considering a steady load

To obtain a physical appreciation of the mechanisms that control power loss in a bearing it is helpful to consider the steady load case shown in Fig 5. This particular analysis has been simplified by assuming that the load carrying film extends throughout the converging part of the clearance space (ie from h_{max} to h_{min} the so called π film extent). The figure shows how power loss (in dimensionless terms) varies with the journal eccentricity ratio.

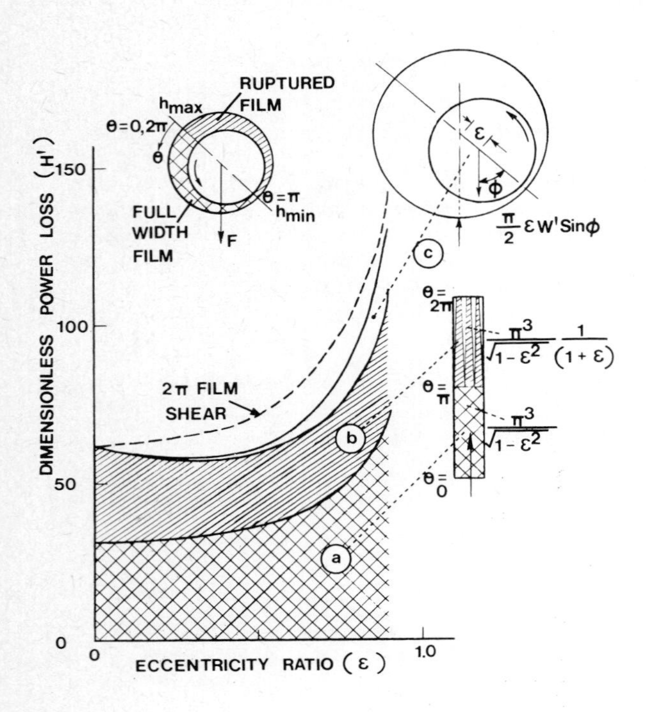

Fig 5 Schematic diagram showing power loss components for steady load conditions

Three effects labelled (a) (b) and (c) in Fig 5 control the power loss. The lower chequered area (labelled (a)) represents the shear losses generated in the full width film condition of the converging film; the dimensionless power loss for this region is a function of the eccentricity ratio:-

$$H' = \pi^2 c_r \int_{\theta=0}^{\theta=\pi} \frac{d\theta}{h} = \frac{\pi^3}{(1-\epsilon^2)^{0.5}} \qquad ..[1]$$

The second component (labelled (b)) is due to shear of the oil in the diverging region. Here a constant volume of oil (controlled by the cross section at h_{min}) is carried around circumferentially, splitting up into streamers; the power loss is dependent upon a modified function of eccentricity ratio.

The component marked (c) is due to the hydrodynamic film pressure (14) and is effectively a torque term; it is a function of journal position and load.

When the journal is concentric with the bearing ($\epsilon = 0$), the angular extent of the oil film will be 2π. The friction equation (14) for this case (sometimes called the Petroff equation) is given by:-

$$F_F = \frac{\text{shear area x viscosity x velocity}}{\text{film thickness}}$$

$$= \frac{2\pi r b \eta U}{c_r}$$

or in terms of power loss

$$H = F_F \omega r$$

$$= \frac{2\pi^3 \eta N^2 b d^3}{C_d}$$

or in terms of dimensionless power loss H' (see notation)

$$H' = 2\pi^3$$

(ie 62.0 on the vertical axis in Fig 5).

If shear effects only were taken, and it was assumed that a complete (non-ruptured) film occurred over the full eccentricity range then the losses shown by the "2π film shear" line would be predicted. It can be seen that oil film rupture lowers the bearing loss.

4.2 Dynamically loaded bearings

In dynamically loaded bearings the situation is much more complex. Whilst the load carrying film also extends just over 180° for a finite bearing solution (apart from film history solution (15)) it is not, however, restricted to the converging wedge part of the film, because the dynamic case also relies on 'squeeze' action for the generation of film pressures. This is illustrated schematically by the various positions of the 180 degree load carrying film (approximate solution) shown in Fig A1 in the appendix. Consequently one cannot use the simple $(1-\epsilon^2)^{-0.5}$ term for friction in the load carrying film of engine bearings as the film is seldom from maximum to minimum film positions. As an expedient, a completely full 2π film is often considered in engine bearings (where there is symmetry of oil film extent about the line of journal and bearing centres). For this case the shear component is simply given as twice that in equation [1].

$$H' = 2\pi^3/(1 - \epsilon^2)^{0.50} \qquad ..[2]$$

The advantage of this equation is its simplicity. However, it does not allow for film rupture in the bearing, and this can be important. Other factors which should be considered when predicting power loss in engine bearings are:-

a) for big end (connecting rod) bearings the angular velocity of the big end bearing ω_b is not constant with time, and this should be considered together with the journal angular velocity ω_j

and b) the actual translatory motion of the journal centre within the bearing clearance space.

4.3 General friction torque and power loss equations for the dynamic load case

One of the most comprehensive equations for predicting power loss in bearings is given by Booker, Goenka and van Leeuwen (16) in vector and cross-product format. The author of this review paper agrees with the validity of this equation and has rearranged it into algebraic form to make it more understandable and so that it can be more readily compared with the work of others. A summary of such equations and their development are given in Tables 1 and 2, for the geometric notation* shown in Fig 6.

Table 1 – Rate of work done on oil film due to **Rotation** of journal and bearing

A + B	A Shear Term	B Pressure Term
Torque τ_j	$\eta r^3 \iint \frac{1}{h}\, d\theta dz \left[\omega_j - \omega_b\right]$	$r \iint \frac{h}{2} \frac{\partial P}{\partial \theta}\, d\theta dz$
τ_b	$-\eta r^3 \iint \frac{1}{h}\, d\theta dz \left[\omega_j - \omega_b\right]$	$r \iint \frac{h}{2} \frac{\partial P}{\partial \theta}\, d\theta dz$
Power Loss $\omega_j \tau_j$	$\eta \frac{r^3 b}{c} J_1^{00} \left[\omega_j - \omega_b\right] \omega_j$	$\omega_j F_j \frac{e \sin \phi}{2}$
$\omega_b \tau_b$	$-\eta \frac{r^3 b}{c} J_1^{00} \left[\omega_j - \omega_b\right] \omega_b$	$\omega_b F_j \frac{e \sin \phi}{2}$
$H_{ROT} = \omega_j \tau_j + \omega_b \tau_b$	$\eta \frac{r^3 b}{c} J_1^{00} \left[\omega_j - \omega_b\right]^2$	$\left[\frac{\omega_j + \omega_b}{2}\right] F_j e \sin \phi$

Table 2 – Rate of work done on oil film due to **Translation** of journal centre O_j with respect to bearing centre O_b

C	Translatory Term C
Power Loss	
$H_{Trans} = \underline{F}_j \cdot \underline{V}$	$\underline{F}_j \cdot \underline{V} = \lvert \underline{F}_j \rvert \lvert \underline{V} \rvert \cos \beta$
where $\underline{V} = \left.\frac{de}{dt}\right\rvert_{x,y}$	$= F_j V \cos \beta$

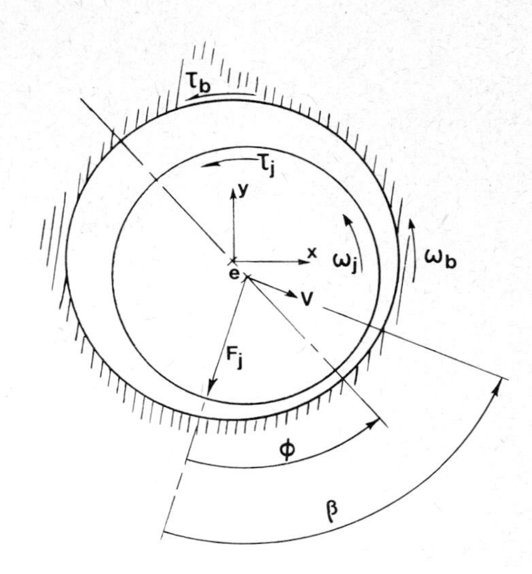

Fig 6 Notation for power loss equations in tables 1 & 2

Table 1 relates to the rate of work done on the oil film due to **rotation** of journal and bearing, resulting in a shear term A and a pressure term B. Table 2 relates to the work done on the oil film due to **translation** of the journal centre (term C).

The resulting power loss at any instant is given by:-

$$H = H_{Rot} + H_{Trans}$$
$$= \left[\eta \frac{r^3 b}{c} J_1^{00} (\omega_j - \omega_b)^2 + \frac{(\omega_j + \omega_b)}{2} F_j e \sin \phi\right] + \left[F_j V \cos \beta\right] \quad ....[3]$$

For a 180° load carrying film (such as given by short bearing theory)

$$J_1^{00} \text{ will be equal to } \int_{\theta_1}^{\theta_1 + \pi} \frac{d\theta}{(1 + \epsilon \cos\theta)}$$

Values for this integration can be found in Booker's table of journal bearing integrals (17). The detailed form of the integration will depend on the attitude angle ϕ and whether the eccentricity is decreasing or increasing, as shown in the Appendix (Fig A1). For a completely full film (no film rupture) J_1^{00} is simply

$$2\pi/(1 - \epsilon^2)^{0.5}$$

The power loss equation [3] gives a good basic foundation for further analysis and for comparing the work of others.

*Footnote – Both Booker and Martin define ϕ positive in a counter clockwise direction. However, Booker considers ϕ from $\underline{\epsilon}$ to $\underline{F}$ (as $\underline{\epsilon}$ is used as a reference axis) and Martin considers ϕ from $\underline{F}_j$ to $\underline{\epsilon}$ (to be consistent with the special case of a steadily loaded bearing). This results in $\phi_{Martin} = 2\pi - \phi_{Booker} = -\phi_{Booker}$ and is instrumental in transforming Booker's equation to those in Tables 1 and 2 as summarised in Table 3.

4.4 Different forms of prediction equations

Several terms make up the general power loss equation [3] and these can be made more applicable to particular situations by choosing the appropriate frame of reference. The friction losses due to shear will be the same irrespective of where an observer is placed, since the shear term involves the **difference** of two angular velocities. The total power loss is not affected by the choice of the reference frame and therefore the **sum** of the remaining terms, labelled 'pressure' and 'translation', will be unchanged, although they will vary individually depending on the motion of the observer (ie on the reference frame).

Such changes to the 'pressure' and 'translation' terms can be illustrated by an example case where the 'observer' is rotating at an average film velocity of $(\omega_j + \omega_b)/2$. For this case the 'pressure' term goes to zero and the 'translation' term will be the product of the force from the journal acting on the oil film F_j and the apparent velocity of the journal V_ξ parallel to the instantaneous load line. The resulting equation for this case is :-

$$\text{Power Loss} = \frac{\eta r^3 b}{c} J_1^{00} (\omega_j - \omega_b)^2 + F_j V_\xi \qquad \ldots\ldots[4]$$

This form of equation is particularly useful for rapid solutions (eg employing the Mobility concept (18)) as V_ξ is directly related to the Mobility Vector M_ξ by

$$V_\xi = \frac{F\left(\frac{c}{r}\right)^2 c}{\eta\, bd} M_\xi$$

and for short bearing theory (18)

$$M_\xi = \frac{(1 - \varepsilon\cos\emptyset)^{5/2}}{\pi} \left(\frac{d}{b}\right)^2 \qquad \ldots\ldots[5]$$

There is a wide range of prediction equations (based on both theory and experiment), for frictional torque, frictional mean effective pressure and power loss, to be found in the literature (2) (8) (16) (19) (20) (21). Many different styles of equation from various sources are shown in Table 3, differing in both format and result.

Table 3 - Power Loss, Friction and Friction Mean Effective Pressure Equations

SOURCE	FORM OF EQUATION	REMARKS
USA BOOKER GOENKA van LEEUWEN (16)	$H = \frac{\eta r^3 b}{c} J_1^{00} \Delta\underline{\omega}\cdot\Delta\underline{\omega} - \underline{e}\times\underline{F}_1\cdot\bar{\underline{\omega}} + \underline{F}_j\cdot\dot{\underline{e}}$ $= \frac{\eta r^3 b}{c} J_1^{00} \lvert\Delta\underline{\omega}\rvert^2 - \lvert\underline{e}\rvert\,\lvert\underline{F}_1\rvert\,\lvert\bar{\underline{\omega}}\rvert\sin\phi + \underline{F}_1\cdot\dot{\underline{e}}$	Basis for Table 1 & 2
UK TABLE 1 & 2 (This paper)	$H = \frac{\eta r^3 b}{c} J_1^{00} (\omega_j - \omega_b)^2 + e F_j \sin\phi \frac{\omega_j + \omega_b}{2} + F_j V \cos\beta$	$\phi_{Martin} = -\phi_{Booker}$ Explains sign change of pressure term
AUSTRIA GROBUSCHEK and EDERER (19)	$N_R(\alpha) = \left[\frac{\eta BD}{\psi}(\omega_W - \omega_S)\frac{\pi}{\sqrt{1-\epsilon^2}} + \frac{\psi\epsilon}{2} P \sin(\delta - \gamma)\right] \times \frac{D}{2}(\omega_W - \omega_S)$ $H = \frac{\eta r^3 b}{c} J_1^{00} (\omega_j - \omega_b)^2 + e F_j \sin\phi \frac{\omega_j - \omega_b}{2}$	Pressure term differs from Table 1. Believed to be incomplete - see appendix
JAPAN SOMEYA (20)	$N_R^*(\alpha) = N_R(\alpha) + N_R'(\alpha)$ $= \int^B\int^{2\pi}\left(\frac{h}{2}\frac{\partial P}{\partial\phi} + \eta\frac{U_W - U_S}{h}\frac{D}{2}\right) dz\,d\phi\,\frac{D}{2}\omega_W$ $+ \int^B\int^{2\pi}\left(-\frac{h}{2}\frac{\partial P}{\partial\phi} + \eta\frac{U_W - U_S}{h}\frac{D}{2}\right) dz\,d\phi\,\frac{D}{2}\omega_S - PV\cos\lambda$	Unclear why the term involving ω_s (shell angular velocity) has positive sign
JAPAN NAGAO et al (2)	Friction losses:- crankshaft system $P_{f(c)} \propto \eta L^{0.7} D^3 N C^{-0.5}$	From motored engine
USA BISHOP (8)	$MEP = \frac{B}{S}\frac{N}{1000}\frac{6}{B^3}(a^2 c + b^2 d/m + e^2 f)$	Empirical and from motored engines
UK SPIKES and ROBINSON (21)	$\text{POWER LOSS} = \frac{\eta r^3 \ell \omega^2}{c}\frac{\pi}{\sqrt{1-\epsilon^2}}\left(2 + \frac{3\epsilon^2}{2+\epsilon^2}\right)$	Correlates with experiment (21) (13). ϵ = average eccentricity ratio though

4.5 Method incorporating environmental factors

When studying friction in connecting rod big end bearings three components of load are considered, from reciprocating inertia forces, rotating inertia forces and gas forces due to firing. The reciprocating mass at the small end (consisting of piston, piston pin and part of the connecting rod) together with the rotating mass (remainder of the connecting rod mass) produce the inertia forces acting on the big end bearing. The remaining component is the firing load. Whilst the magnitude of the firing load does not affect the friction very much (a factor of two on firing load may change the friction by only 10 to 20%), it is important that the firing load be considered. The reason for this is that the friction is influenced by the characteristic shape of the journal centre orbit, which is certainly very much modified if one neglects the firing load completely. Martin, Booker and Lo (22) studied the influence of engine inertia forces on power loss in connecting rod bearings and found that the inertia components of load alone were not sufficiently descriptive for power loss prediction.

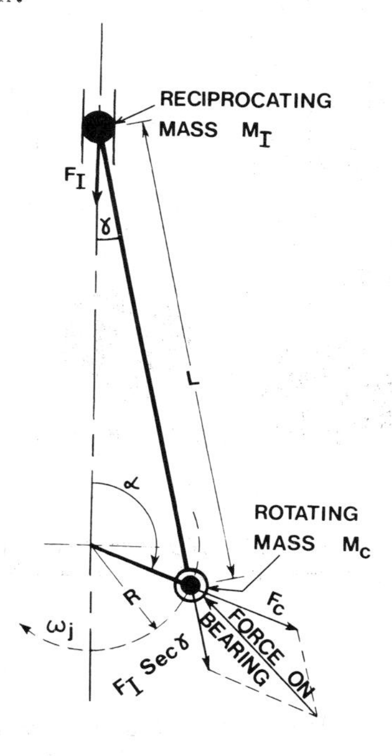

Fig 7 Reciprocating and rotating mass system

Power loss prediction charts are presented in a very general form for a range of bearing length to diameter ratios (b/d from 0.4 to 0.6) in Figs 8 and 9. These charts can be used outside this range (for b/d from 0.2 to 0.7) with little loss in accuracy. The characteristic shapes * of most big end bearing load diagrams have been allowed for by defining them in terms of an inertia shape factor B/A and a firing load factor C/B.

*Footnote – In this study the polar load diagrams are relative to the connecting rod axis and have an 'elliptical' inertia loop; a pear shaped inertia loop is obtained when the load diagram is plotted relative to the cylinder axis (6).

The values A, B and C are dependent on the reciprocating mass at the small end of the connecting rod M_I (ie the mass of the piston assembly (Fig 7) and part of the rod), the rotating component of mass M_c at the big end and the maximum cylinder pressure P_{cyl}.

The load diagram shape ratios can be calculated from:-

$$B/A = (M_I/M_c) + 1$$

$$\text{and } C/B = (C/A) \times (A/B)$$

$$= P_{cyl}\, A_{cyl}/(2M_c\omega_j^2\, R\, ((M_I/M_c) + 1))$$

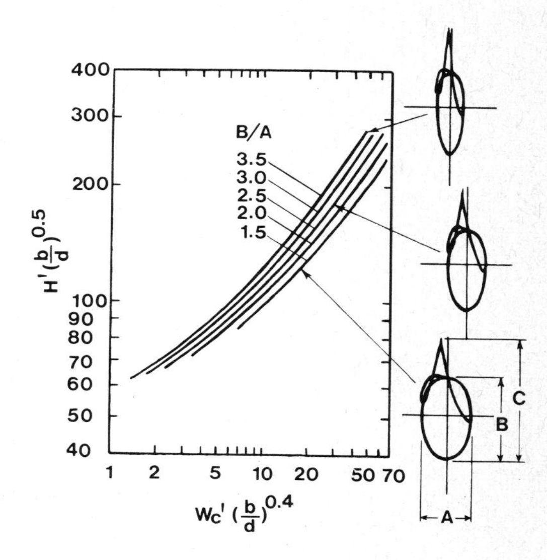

Fig 8 Prediction of power loss in connecting rod big-end bearings; b/d from 0.4 to 0.6; B/A variable; C/B equal to 1.5 (see Fig. 9 for other values of C/B)

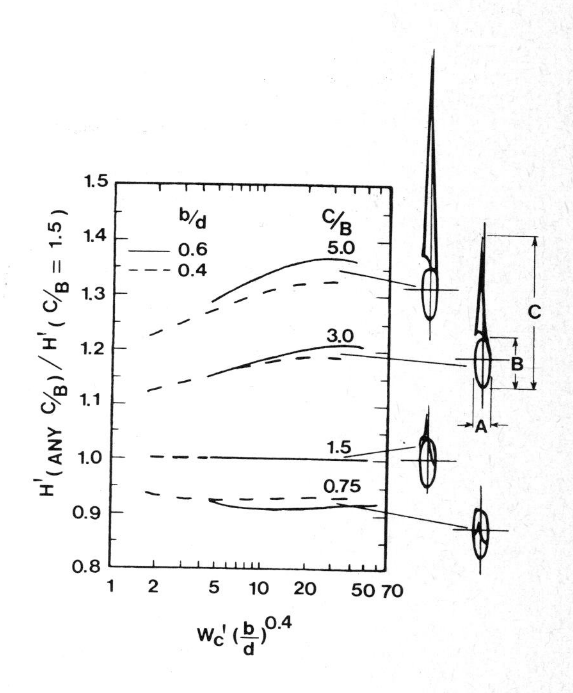

Fig 9 Power loss modifying factor for different C/B ratios b/d from 0.4 to 0.6; C/B variable; B/A from 1.5 to 3.5

Two prediction charts, Fig 8 and Fig 9, have been produced to enable power loss to be estimated. The load diagram shape is represented by the various curves and the load diagram size (using the rotating load component F_c) together with the bearing geometry is represented by the dimensionless load number W'_c. The charts are based on a finite bearing solution and several hundred cases have been computed, for different load diagram shapes and b/d ratios,to arrive at these curves. The dimensionless term H' represents average values of the cyclic variations in power loss.

The b/d terms raised to the power of 0.5 and 0.4 bring all the separate studies for various b/d ratios together. In the same way the power loss modifying factor in Fig 9 corrects the results for different firing loads. As well as giving an immediate assessment of power loss, the design charts are also useful in showing the effect of altering engine and bearing variables (such as piston mass, speed, bearing length, clearance, viscosity etc). Detailed trends are further discussed in a later section of this paper.

5. PREDICTED BEARING FRICTION LOSS FOR THE VEB STUDY CASE

The connecting rod bearing for the Ruston and Hornsby 600 HP, 6 cylinder 600 rev/min VEB MK III diesel engine must be one of the most analysed bearings in the world. The Campbell et al review in 1967 (5) which first used this bearing was mainly concerned with experimental and theoretical predictions of oil film thickness. Subsequently oil flow (6), film pressure (6), (23) and temperature (24) have been considered. However, it is only recently that power loss associated with this particular bearing study has appeared in the literature (25).

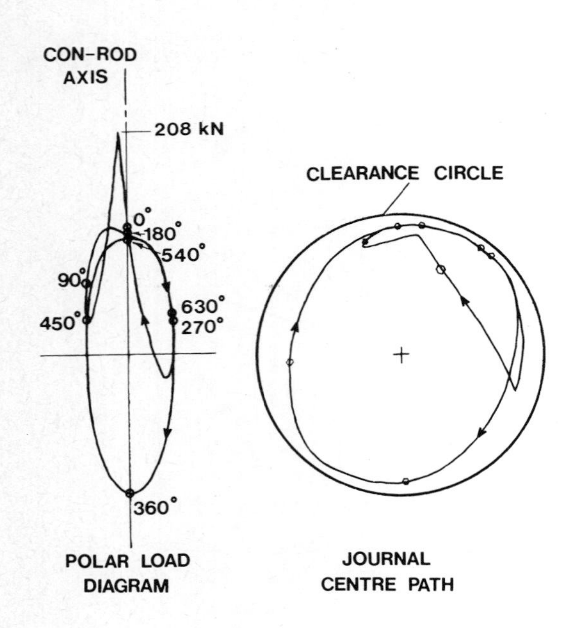

Fig 10 Polar load diagram and journal centre path for VEB connecting rod big-end bearing

5.1 Accumulative factors of friction (hydrodynamic)

The various factors of friction can be specifically defined to match the mathematical terms described in tables 1 and 2. Such terms for the VEB bearing example have recently been computed, for this review, by both the Glacier Metal Company and General Motors Research Laboratories, giving very similar results. A finite bearing mathematical model was used taking into account the oil supply pressure and how it affected the oil film extent (note that this is not the 'film history' case referred to later).

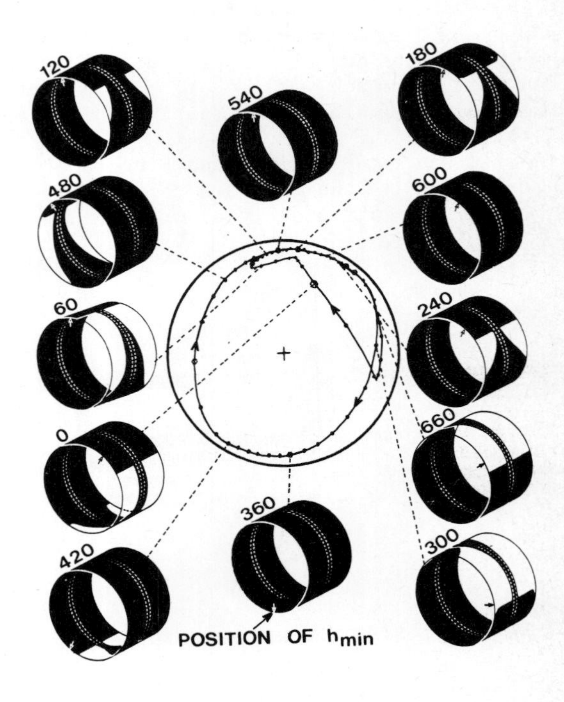

Fig 11 Film extent for VEB study case
Data from General Motors Research Laboratories (26)

Details of the VEB bearing and relevant engine data are given in the Appendix of ref (5). Fig 10 here shows the polar load diagram and a typical journal centre orbit path consistent with previous published results (5) (6). The outer part of Fig 11 shows 'thumb nail' sketches of the oil film extent at every 60 degrees of crank rotation as predicted by General Motors Research Laboratories (26). The circumferential dotted lines represent a full circumferential oil groove, the black area (either side of this) is the predicted film shape (under pressure) and the small arrows indicate the position of minimum oil film thickness for the various crank angle positions. The film shape and film position vary considerably throughout the 720 degrees of crank rotation. At 360, 540 and 600 degrees the bearing is completely full of oil as can be seen from the distribution of the power loss components shown in Fig 12. In this figure (computed by GM) the significance of the various 'mathematical' terms of tables 1 and 2 can be be assessed.

The numbered areas in Fig 12 showing the bearing power loss distribution are related to the power loss terms in Tables 1 and 2 as follows:-

	Fig 12 code	Table 1 and 2
Pressure effect	1	Term B
Translation effect	2	Term C
Shear effect (film as Fig 11)	3	Term A
Shear effect (2π film)	3 + 4	Term A

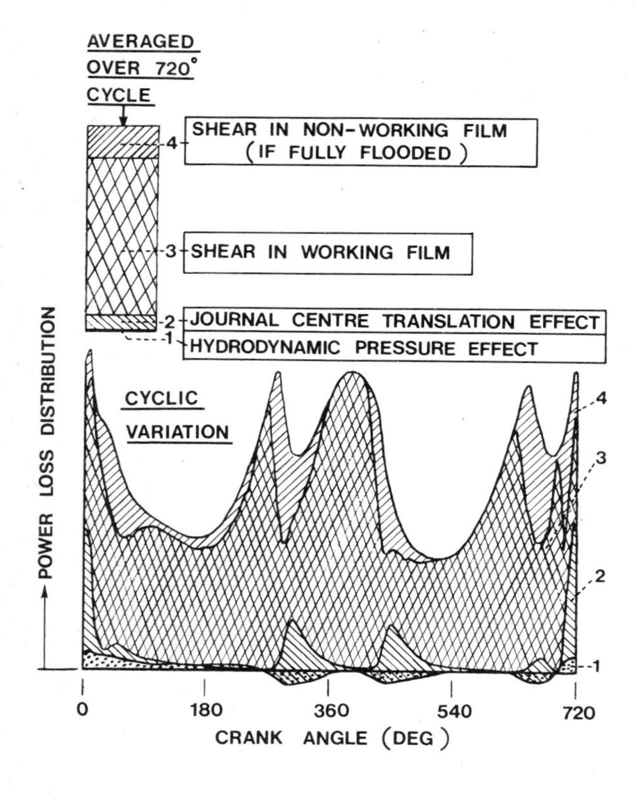

Fig 12 Distribution of bearing power loss for VEB Study case — computed by General Motors Research Laboratories (26)

The hydrodynamic pressure term taken by itself can be negative as well as positive, depending on whether the minimum film thickness position is in advance of or lagging behind the working film. This can be seen in Fig 11 where the minimum film position (small arrow) is in advance at 60° crank angle and lagging at 300° crank angle. This is consistent with the positive and negative values due to the hydrodynamic pressure effect, at these particular crank angles in Fig 12 (shown by the dotted areas). As mentioned earlier the sum of the pressure term and the translation term is independent of the position of the 'observer'. These two terms added together (or the $F_j V_s$ term of equation 4) will always be positive.

The journal centre translation effect is seen, from Fig 12 (area marked 2), to be significant at zero degrees crank angle and this corresponds to the journal centre moving with a high velocity, as shown by the wide spacing of the 10° crank interval markers on the orbit in Fig 11. The shear in the working film is represented by the chequered area in Fig 12 and the total power loss at each individual crank position is given by the summation of the areas marked 1, 2 and 3 (ie the top of the chequered area from the datum) in Fig 12. If instead of having the film extents of Fig 11 the bearing clearance was fully flooded, then an additional component 4 would have to be added to the total. The top line therefore represents the shear effect for a bearing with a 2π circumferential film extent together with the pressure and translation effects. The average values over the full 720° cycle for each of the individual 'components' are shown at the top of Fig 12.

5.2 Comparison of power loss from various sources

Power loss predictions from various sources, for the VEB study case, are compared in Fig 13 and Fig 14a & b. Fig 13 shows the effect of oil film extents using results from various finite bearing solutions. Case (a) neglects the feed pressure and the working film extent is usually just slightly greater than 180 degrees. (Moes and Bosma (27) give a graphical guide for the film extent in dynamically loaded bearings and Goenka has presented curve fit equations (23)). Case (c) is for the theoretical case of a complete film over an angular extent of 2π, including pressure and translatory effects as well as shear. Various computed solutions including the General Motors finite element method (25), Glacier finite difference method (6), and Glacier using Moes

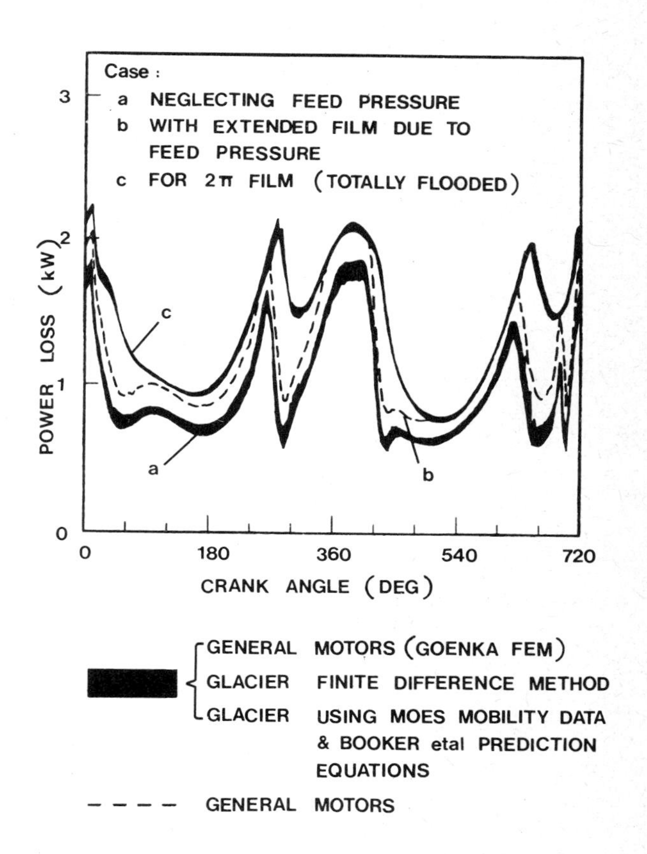

Fig 13 The effects of various film extents on power loss (VEB)

finite bearing mobility data (18) with Booker's Mobility method (18) (5), all lie within the fairly narrow black bands labelled (a) and (c). The intermediate case, curve (b), is from the General Motors results and includes supply pressure effects (film extents as Fig 11).

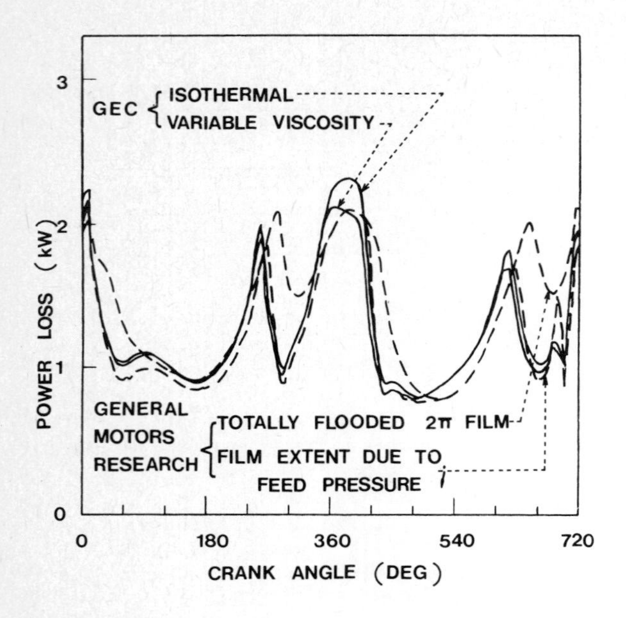

Fig 14a

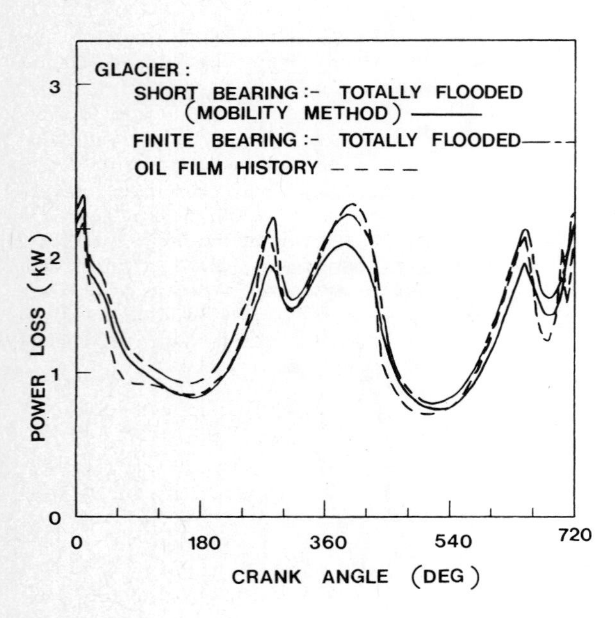

Fig 14b

Fig. 14 a & b—Cyclic variation in power loss
—VEB Study case results from different sources

GEC (UK) have also kindly carried out studies for this paper using their optimised short bearing solution (28). This has improved accuracy over the standard short bearing theory especially at high eccentricities and therefore should be more comparable with finite bearing solutions. Unique speed and boundary parameters Ω and α (in place of ϵ, $\dot{\epsilon}$, ω and $\dot{\phi}$ etc) form basic terms in their Reynolds equation. Their equations for power loss can be transformed into a similar form to that shown in tables 1 and 2. GEC's predictions for power loss (29) in Fig 14(a) include the use of a heat balance with variable viscosity (with time); they have also predicted the isothermal case. Furthermore they have attempted (albeit approximately) to take some account of shear in a zero pressure striated film region. Their predictions of power loss for the VEB case are compared with the General Motors result, which are shown by dashed lines in Fig 14a. The GEC cases, with the striated model, would have been expected to be within the General Motors dashed lines, had the eccentricities and viscosities been similar. However with the different models considered the trends are as expected.

Table 4 – PREDICTED POWER LOSS (including hydrodynamic pressure and translation effects) for VEB STUDY CASE

SHEAR FILM EXTENT	TOTAL POWER LOSS kW: LOAD CARRYING FILM (NEGLECTING FEED PRESSURE)	CONSIDERING FEED PRESSURE	WITH STRIATED FILM INCLUDED	TOTALLY FLOODED (2π FILM EXTENT)
FINITE BEARING - ISOTHERMAL				
GENERAL MOTORS (FINITE ELEMENT)	1.02	1.19	-	1.39
GLACIER (FINITE DIFFERENCE)	0.97	1.09	-	1.375
GLACIER (MOES MOBILITY MAP)	0.94	-	-	1.39
GEC (OPT. SHORT BEARING)	-	-	1.21	-
GLACIER (OIL FILM HISTORY)	-	0.95	1.29	-
FINITE BEARING - VARIABLE VISCOSITY				
GEC	-	-	1.71	-
SHORT BEARING THEORY				
GLACIER (BOOKER'S MOBILITY MAP)	0.84	-	-	1.26

The cyclic variations of power loss for three different theoretical models are compared in Fig 14b. The 2π film (totally flooded) shear, film pressure and translatory effects are all included in the short and finite bearing models. However, the load carrying film extent is π for the short bearing and slightly greater than π for the finite bearing solution. The finite bearing solution gives higher power losses (chain dashed lines) than the short bearing solution (full line) because of the smaller oil film thickness involved in the former. However the trends are very similar. The third curve (dashed line) is based on a so called 'oil film history' model (15) (6). This involves the study of oil transport within the bearing oil film and takes into account the effects of insufficient oil to fill the load carrying area of the bearing. Such a theoretical model, which has reduced film extents, has been found to have fairly good correlation with practice, especially for oil flow (6). The reduction in friction that would be gained due to reduced film extent, however, may be offset by increased friction due to resulting smaller oil films. This generally explains why all the curves in Fig 14b have similar trends. Because of the difficulty of defining the shear condition within the striated region, the shear over a complete full width film has been considered; this will predict the upper limit to power loss. The actual shear area may be only slightly less than this for the VEB bearing, since the oil is supplied from a circumferential groove. However, with the more common practice of an ungrooved big end bearing, fed from a crankpin drilling, one would expect less filling of the bearing clearance space. Under such conditions the oil film history model would be more representative of actual performance than the other solutions.

A summary of predicted power loss for the VEB study case is given in table 4, and includes average values for the cases shown in figures 13 and 14a and b from General Motors Research Laboratories, GEC and the Glacier Metal Co.

6. HOW DIFFERENT VARIABLES AFFECT BEARING FRICTION

6.1 Mass of moving parts

The mass of the moving parts in an engine create inertia forces which generally increase friction. However, the reduction of masses is not a simple route to the reduction of friction, as this has to be considered in conjunction with component strength and reliability of performance at the friction producing surfaces. Also vibration and noise may be affected by change in component mass.

The effect of connecting rod mass on the big end bearing power loss can be seen in Figs 8 and 9 for particular cases. The mass system involving the rotating component M_c at the big end and the reciprocating component M_I at the small end is allowed for in the terms B/A, C/B and W'_c. The power loss is mostly affected by the load number W'_c, the b/d ratio and the shape of the inertia loop in the load diagram defined by B/A. Fig 8 involving all of these variables can be used separately (from Fig 9) to obtain a reasonable estimate of power loss. The modifying factor on power loss due to various firing loads, Fig 9, is less significant but nevertheless still important when the value of C/B is high.

The reciprocating mass M_I includes the piston assembly and any weight reduction here could result in lower friction, both at the piston/cylinder interface and at the big end bearing. There will be greater total benefit at the piston, as piston losses form a major part of the mechanical losses. Many designs of lightweight piston have been developed. One novel design developed by Bruni at A E BORGO Spa, Italy (30) is the 'x' piston* shown in Fig 15, which has much of the piston skirt removed forming an X shaped support. Such a design will have a small but beneficial effect on the big end bearing friction (due to reduction of B/A ratio in Fig 8).

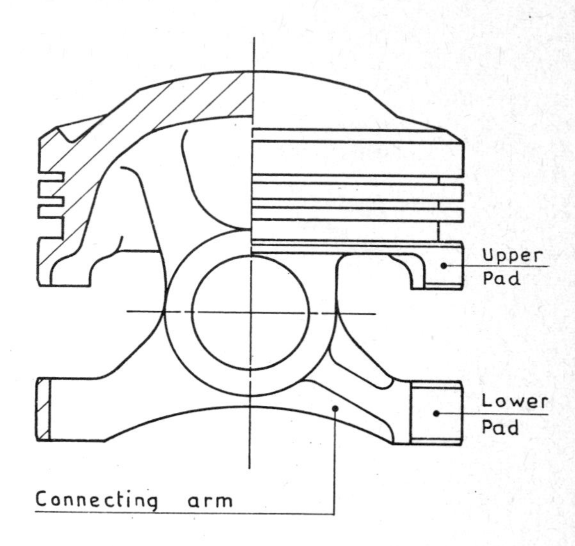

Fig 15 Reduced weight 'x' piston
AE Borgo Spa — Bruni (30)

The effect of reducing reciprocating and/or rotating masses on the connecting rod system will generally result in lower power loss in the connecting rod bearings. Whilst the resultant forces will also change on main bearings these may not necessarily all have a corresponding reduction in friction. In V engines changes in connecting rod mass may require modified balance arrangements which in turn will affect the friction. With all these effects to consider, and engine reliability of prime importance, one cannot be certain of improving the friction in main bearings by changing mass.

6.2 The effect of bearing geometry

The effect of bearing size, operating viscosity and speed on power loss in connecting rod big end bearings can be obtained from Fig 8.

For a quick overall view of trends, however, the author has developed a friction torque trend analyser, Fig 16; this has been derived from the slopes of the B/A curves in Fig 8. In Fig 16 one can see at a glance the likely change in friction torque resulting from a change in any of the variables shown. For instance the example dotted lines show that if the diameter is increased by 50% (a multiplication factor of 1.5), then the resulting frictional torque will be increased by a multiplying factor of between 2 and 2.6. Likewise decreasing the diameter will reduce the friction torque. If two variables are changed simultaneously then the multiplying factor on friction torque will be the product of the two resulting individual factors.

A similar friction torque trend analyser, Fig 17, has been developed for main bearings, by considering measured bearing friction from six

*Footnote - British Patent Application 82.21993 and foreign applications pending.

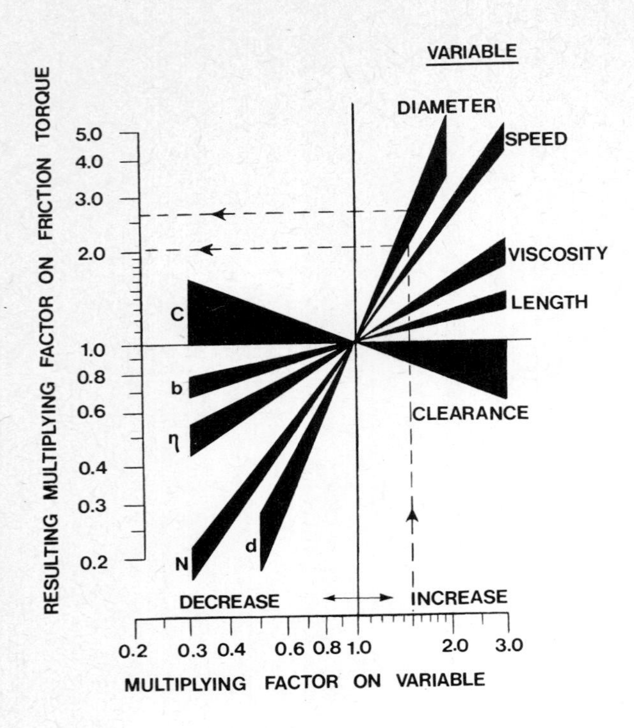

Fig 16 Friction torque trend analyser for connecting rod big-end bearings (predicted)

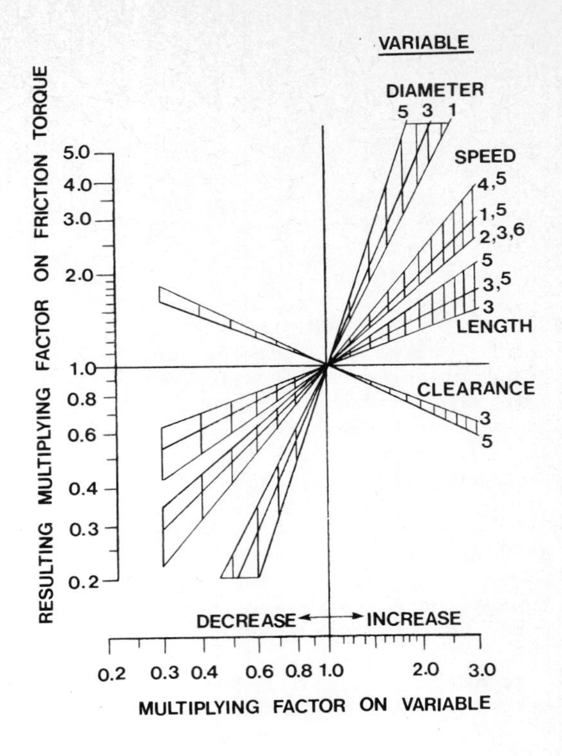

Fig 17 Friction torque trend analyser for main bearings (data from motored engines)

different sources. The radial lines in Fig 17 show the various trends found (with the source identified by the numbers 1 to 6). The data source used and the corresponding literature references are shown below.

Source Reference in Fig 17	Engine - Size	Literature Reference
1	Ford-large range	(8) Bishop'64
2	Vauxhall 1250cc	(13) Vickery'75
3	Mazda 1500cc	(31) Miyachika'80
4	Ford (4 cylinder)	(4) Kovach et al'82
5	Magnum 2000cc	(2) Nagao et al'83
6	Subaru 1300cc	(3) Hoshi'84

It is reassuring that both Fig 16 from theory and Fig 17 from practice (motored engines) have similar characteristic trends (albeit that one relates to big end bearings and the other to main bearings).

Both figure 16 and 17 generally indicate that less friction may occur with a smaller bearing diameter, smaller bearing length, possibly larger clearance and lower viscosity grade of oil. However, these beneficial changes only apply to a bearing operating under hydrodynamic conditions. Reduced bearing size will result in smaller oil film thicknesses and these in turn may place the bearing under mixed lubrication conditions. Under such very thin films the contact area of the surface asperities partly supports the load (32) and there is a tendency towards increased losses. If a reduced size (ie smaller diameter) leads to greater misalignment (in a main bearing) then the increased edge loading effects may also increase total losses as well as perhaps adversely affecting reliability.

At General Motors Research Laboratories, Goenka (25) predicted the effect on power loss of various grooving arrangements, different bearing circumferential shapes and axial film variations as shown in Fig 18. This analysis again relates to the VEB study case and shows the power loss for the full groove case as 1.38 kW (consistent with Table 4). The power loss for the non-grooved case is considerably lower (1.148 kW) as this bearing will be operating with larger oil film thicknesses. This results from the developed hydrodynamic pressure extending axially over the full length of the bearing, whereas for the full circumferential groove case the pressure is only developed on each individual land. The journal centre orbit in Fig 10 shows that most of the friction is developed in the top (rod half) of the bearing; moving three times into small film regions in the rod half and only once in the cap half with a comparable small film. Thus the predicted power loss where there is a half groove (180°) in the cap is not all that different from that obtained with a non-grooved cap. Similarly a 180° groove in the rod half hardly changes the power loss from the full groove case. A single value of operating viscosity (from 5), was used in these particular studies. However, the geometric variations shown in Fig 18 may well affect operating viscosity, and hence friction. In this study, the lowest values of friction are predicted for bearings with no grooves in the rod half. An extension to this particular study, including a heat balance, could prove useful in the future.

Goenka (25) is probably one of the first researchers to predict the performance of different bearing shapes under dynamic loading. In Fig 18(b) the minimum clearances are the same for the various profiles. The differences in power loss for the various bearing shapes are not very great and the order of merit based on friction could possibly alter if a heat balance influenced by these shapes were included. The axial variations in Fig 18(c) again show little difference in power loss. The amount of taper and misalignment considered is 1 in 10 000 per unit length, and the diameter of the shaft in the 'hour glass' arrangement is 0.005% smaller at the centre than at the ends.

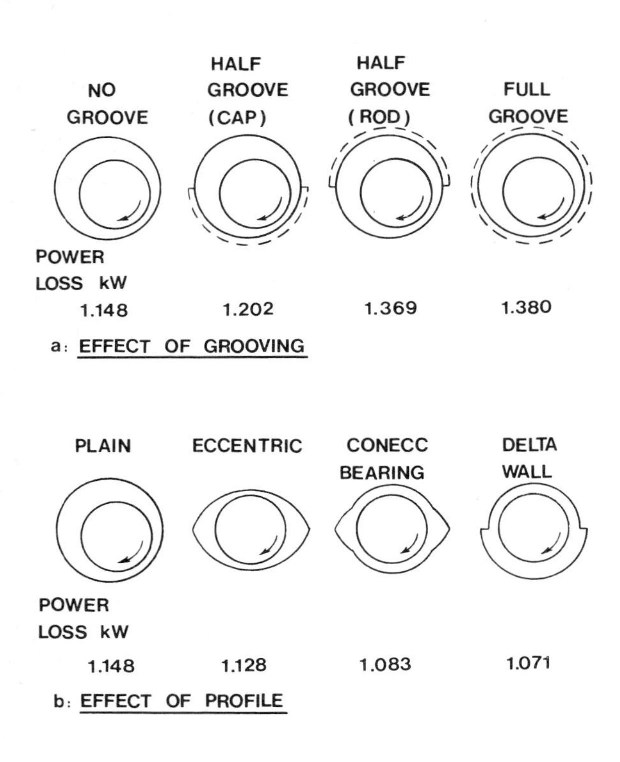

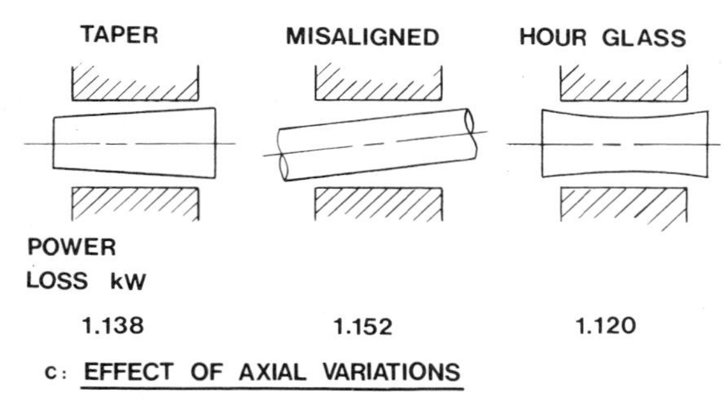

Fig 18 The effect of bearing geometry on power loss for the VEB study case

6.3 Reduced friction by re-design

A significant reduction in friction can sometimes be achieved in practice by re-design, as shown by the following work in Japan.

The Nissan Motor Company (1) designed a new engine, designated CA20, for transverse location in a front wheel drive vehicle. In doing this they compared performance with an earlier design, designated Z20, of similar capacity (but not suitable for transverse location). The engine is a 2 litre, 4 cylinder petrol engine and the transverse design involved weight reductions as well as different bearing sizes. They set out to achieve a light weight compact size engine, with

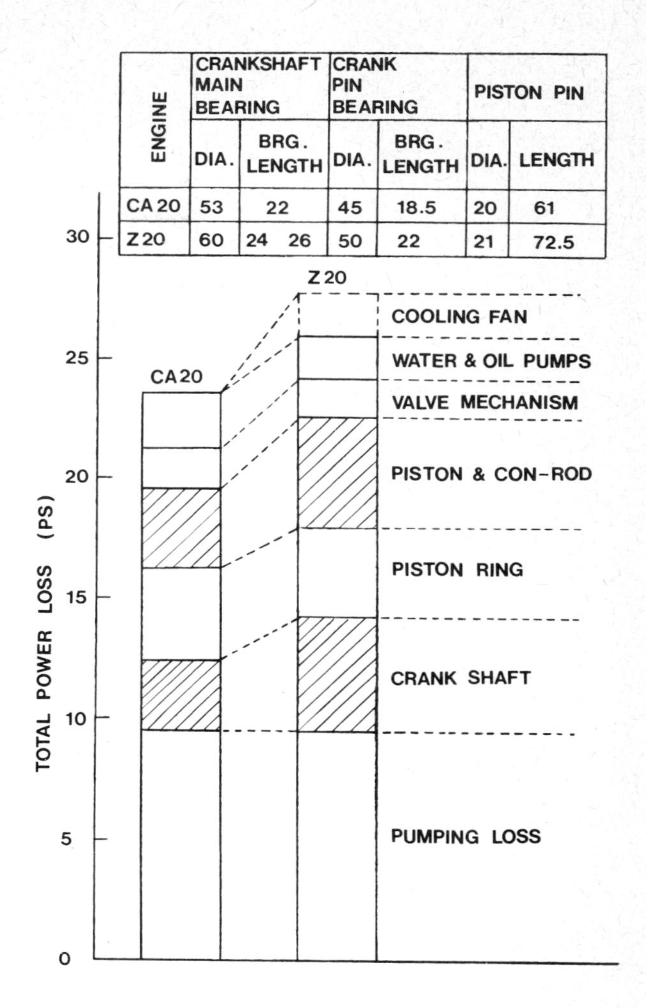

ENGINE	CRANKSHAFT MAIN BEARING		CRANK PIN BEARING		PISTON PIN	
	DIA.	BRG. LENGTH	DIA.	BRG. LENGTH	DIA.	LENGTH
CA20	53	22	45	18.5	20	61
Z20	60	24 26	50	22	21	72.5

Fig 19 Friction loss — Nissan Motor Company (1)

fuel economy amongst other objectives, stating that these were attained almost to the maximum extent. This study therefore illustrates the extent of possible changes. Going from the Z20 to CA20 engine, the percentage weight reductions amounted to 11% on the piston assembly, 32% on the crankshaft, and 29% on the connecting rod, with other reductions throughout the engine. The change in bearing dimensions and friction loss (1) is reproduced in Fig 19. In this figure the lower friction loss for the CA20 engine is immediately evident especially for the main (crankshaft) bearings and to some extent for the big end bearing (although this is classified with piston losses as well).

Also in Japan, Hoshi presented experimental results on friction losses in engines (3) and discussed ways of improving fuel economy by reducing such losses. By applying the results to an engine for Subaru cars, Hoshi found that a reduction of 17 to 21% of the total friction losses became possible, amounting to improvements in fuel economy of 7% on city roads and 3% on highways. The largest reduction in friction was associated with the piston system (piston friction reduced by a quarter) amounting to a saving of about 9 to 11.5% of the total friction losses. Compared to this, a practical change in bearing dimensions (reducing diameter and length) achieved a saving of 3% of the total friction loss.

7. COEFFICIENT OF FRICTION

The coefficient of friction is sometimes presented in a form similar to Fig 20 (often referred to in dimensional terms as a Stribeck curve) showing hydrodynamic, mixed and boundary lubrication regions. The effect of reducing bearing length and diameter (moving from right to left on this figure) is to decrease friction coefficient when at A in the hydrodynamic region and increase friction coefficient when at B in the mixed lubrication region. For lowest friction one would like to operate at C. When fully hydrodynamic but with small oil films (as at A), there may be a danger of scoring of the bearing surface if there is inadequate filtration of the oil or if the surface roughnesses of the bearing surfaces are relatively large. To the left of A on the curve there could be metal wiping due to high temperatures produced from asperity contacts and the possibility of seizure in the high friction region.

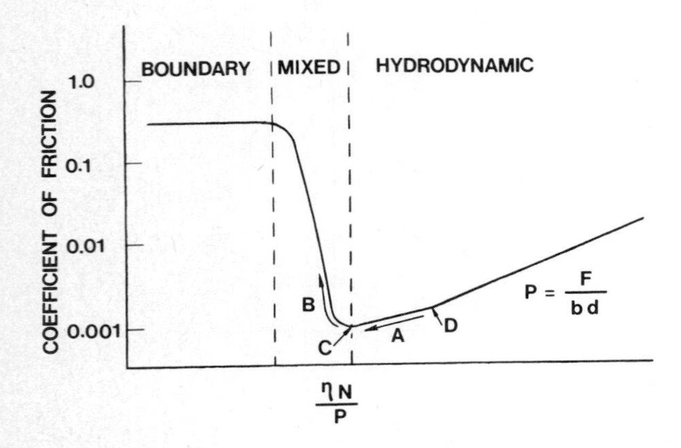

Fig 20 Schematic representation of the various lubrication regions

The curve in the boundary region would depend on the material properties of the two mating surfaces (including hardness). In the mixed lubrication region the coefficient of friction will depend on the surface roughness of the surfaces in contact (taking account of any prior 'running in') (33) (34). In the hydrodynamic region a broad band of curves would have to replace the single curve if various b/d and c_d/d ratios were to be considered. Vogelpohl shows such a band of results when considering various b/d ratios for steadily loaded bearings (35) and also shows a 'knee' in the curve (similar to the author's point D in Fig 18) and suggests that at this point there is a transition speed, below which surface rubbing may occur with resulting high friction. The author is of the opinion that the transition point where rubbing of the surfaces is likely to occur will possibly be at the ill-defined point C, and point D may in some cases, be well into the hydrodynamic region.

The predicted coefficient of friction for big end connecting rod bearings showing the effect of clearance ratio is given in Fig 21 (distinctly showing the characteristic 'knee' as found in Vogelpohl's steadily loaded bearing analysis). A coefficient of friction value (in dynamically loaded bearings) is dependent on the actual datum load used in the load cycle. In this case the rotating load F_c is used in both the friction coefficient and the $\eta N/P_c$ term. Note that for a constant rotating mass M_c, P_c is proportional to the square of the speed, so increasing speed will move from right to left in Fig 21. At low speed conditions a larger

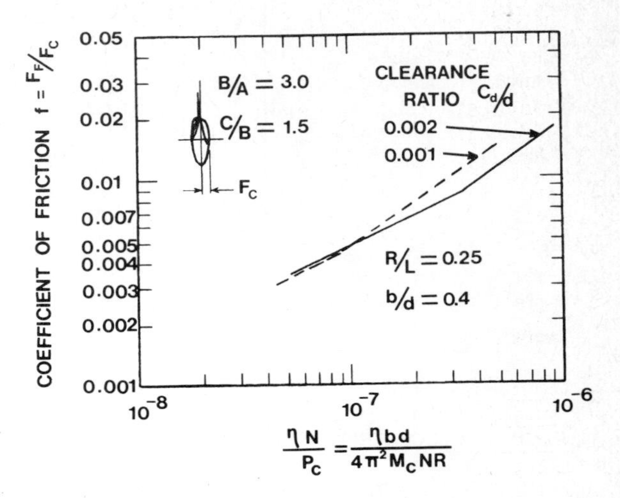

Fig 21 The effect of clearance on coefficient of friction

clearance will reduce friction. However, this could be at the expense of excess oil flow (larger pump requirement) and maybe the danger of bearing attenuated noise. At high speed conditions changing the clearance apparently has little effect on friction (if the viscosity does not change). In practice the viscosity will change with clearance and this re-emphasises the need to carry out a heat balance to obtain a realistic operating viscosity. Clearance increases at high speed may also reduce the oil film thickness, giving greater likelihood of excessive wear.

8 CONCLUDING REMARKS

The results from motored engines show that bearing friction may account for 20 to 30% of the mechanical losses in an engine. Bearing friction is therefore significant and any reduction by careful re-design could help in fuel economy.

To reduce friction in engine bearings, particularly in big end connecting rod bearings it would appear necessary to reduce either the bearing diameter, length, viscosity grade of oil or mass of relevant moving parts. Also an increase in bearing clearance at low speed applications and a decrease in clearances (with consequential lower operating viscosity) at high speed applications may result in reduced friction.

These changes generally tend towards a more adverse situation, however, and great care is necessary in producing a viable design to operate under the required range of conditions. Not only has the bearing to operate reliably, but any new design must meet the requirement of the whole engine system and the various interaction of other engine components.

The prediction charts for power loss and coefficient of friction (Fig 8, 9 and 21) together with the trend analysers (Fig 16 and 17) aid in assessing the effect of different variables on friction.

In aiming for low friction in a bearing there are so many interacting criteria and the chance of bearing damage, that it is imperative that guidance for design is based on a framework of experience together with fundamental concepts. Generally lower friction means lower margins of safety and therefore the designer must have a better understanding of **when** and preferably **why** bearing damage is likely to occur so that it can be avoided by good design (19) (32) (33) (34) (36) (37). Reliable bearing performance is the first prerequisite for good design, economy taking second place if necessary.

9 APPENDIX

9.1 Variation in Film Thickness for the VEB Example with Special Reference to the Friction Integral J_1^{oo}

The integral J_1^{oo} used in the power loss equation 3 and Table 1 takes into account the variation in film thickness over the specified limits θ_1 to θ_2 around the bearing (where θ is measured in the direction of rotation from h_{max}). For short bearing theory the load carrying film will be 180° extent as shown by the black regions in Fig A1, which relates film extent, film thickness variation and position on journal path (relative to load line) for the VEB study case.

a) For a π film extent (generally)

Booker (17) gives: -

$$J_1^{oo} = \int_{\theta_1}^{\theta+\pi} \frac{d\theta}{1+C\cos\theta}$$

$$= \frac{\cos^{-1}(-\delta A) + \cos^{-1}(-\delta B)}{(1-\epsilon^2)^{0.5}}$$

where $A = \dfrac{\epsilon + \cos\theta_1}{1+\epsilon\cos\theta_1}$

$B = \dfrac{\epsilon - \cos\theta_1}{1-\epsilon\cos\theta_1}$

and $\delta = 1,\ \sin\theta_1 \geq 0$

$\delta = -1,\ \sin\theta_1 \leq 0$

From this it can be established that

$\delta = +1$, in the decay region (Fig A1), ie where ϵ increases

$\delta = -1$, in the recovery region, ie where ϵ decreases

b) For a π film from h_{max} to h_{min} position

$$J_1^{oo} = \frac{\pi}{(1-\epsilon^2)^{0.5}}$$

c) For a 2π film

$$J_1^{oo} = \frac{2\pi}{(1-\epsilon^2)^{0.5}}$$

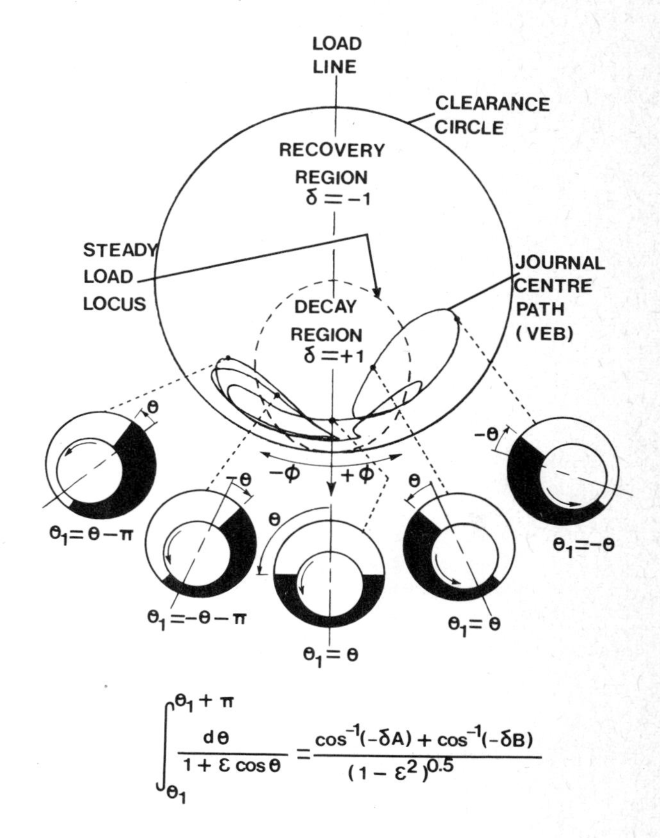

Fig A1 Circumferential film thickness variation at various journal positions (VEB case)

9.2 Comparing Grobuschek's and Ederer's equation with solutions from GEC

Cowking of GEC (private communication) gives the rate of work done on the oil film due to rotation (E_{rot}) and translation (E_{trans}) as:-

$$E_{Total} = E_{rot} + E_{trans}$$

$$= [\{(\omega_j - \omega_b)R\}^2 \iint \frac{\eta\, dxdz}{h} + \frac{(\omega_j + \omega_b)}{2} eF_T] + [\dot{e}F_r - e\dot{\psi}F_T] \quad \ldots\ldots[1A]$$

where F_r and F_T are radial and tangential forces along and at right angles to the line joining the journal and bearing centres and $\dot{\psi}$ is the angular velocity of the line of centres. This is similar to the equations given in tables 1 and 2 and will yield the same result.

Now when observed from a moving member the angular velocities above are replaced by relative angular velocities. In particular consider the frame of reference on the moving bearing.*

Cowking (29) gives $\omega_j' = \omega_j - \omega_b$, $\omega_b' = 0$ and $\dot{\psi}' = \dot{\psi} - \omega_b$ — where the dash (') denotes relative angular velocities

So $E_{Total} = E_{rot} + E_{trans}$

$$= [(\omega_j' R)^2 \iint \frac{\eta\, dxdz}{h} + \frac{\omega_j'}{2} eF_T] + [\dot{e}F_r - eF_T\dot{\psi}']$$

$$= [\{(\omega_j - \omega_b)R\}^2 \iint \frac{\eta\, dxdz}{h} + \frac{(\omega_j - \omega_b)}{2} eF_T] + [\dot{e}F_r - eF_T(\dot{\psi} - \omega_b)] \quad .....[2A]$$

The total rate of work done E_{Total} is the same for both cases irrespective of where the 'observer' is placed, as expected. The pressure term in equation 1A considers the sum of the journal and bearing velocities as Booker et al (16) however in equation 2A the difference between journal and bearing velocities now appear as Grobuschek's equation (19). To compensate for this the translation terms differ by $eF_T\omega_b$. Since Grobuschek and his co author have neglected the translation term, the comparison of the rest of their equation can result in some confusion, as seen in discussions with Someya (20). (See table 3).

*Footnote: It is assumed here that Grobuschek's frame of reference is moving with the bearing, since the minimum film position δ and load vector position γ (in table 3) are measured from the connecting rod axis.

10 ACKNOWLEDGEMENTS

The author wishes to thank the Directors of the Glacier Metal Company for providing time and resources in their Research & Development Organisation, which made the preparation of this paper possible. He would like to thank his colleagues and in particular D R Garner and P M Lo for their help. Also J F Booker of Cornell University for his helpful comments relating to power loss equations.

The co-operation of the following people for providing information for inclusion in this paper is gratefully acknowledged. P K Goenka at General Motors Research Laboratories, Detroit USA; G J Jones at Glacier Metal Co Ltd UK; G S Ritchie and E W Cowking at GEC UK.

Many figures from published work are used in this review and it is hoped that full credit is given by appropriate referencing.

11 REFERENCES

(1) HISATOMI T and HIROSHI L 'Nissan Motor Company's New 2 Liter Four Cylinder Gasoline Engine' SAE paper 820113 SAE Transactions 1982.

(2) NAGAO A, DOINO K, YAMASHITA A, NISHIKAWA T and TANAKA K 'Reduction of Friction Loss for MAGNUM Engine' Mazuda Technical Review 1983. No 1.

(3) HOSHI M 'Reducing Friction Losses in Automobile Engines'. Tribology International Vol 17, No 4 p185-189, August 1984.

(4) KOVACH J T, TSAKRIS E A and WONG L T 'Engine Friction Reduction for Improved Fuel Economy'. Society of Automotive Engineers SAE paper 820085 1982.

(5) CAMPBELL J, LOVE P P, MARTIN F A, and RAFIQUE S O 'Bearings for Reciprocating Machinery: a Review of the Present State of Theoretical Experimental and Service Knowledge'. Proc. I.Mech.E. 1967, 182 part 3A, 51-74.

(6) MARTIN F A 'Developments in Engine Bearing Design' Tribology International June 1983 Vol 16 No 3. Also Proc. of 9th Leeds-Lyon Symposium on Tribology 1982 (Butterworths 1983).

(7) WHITEHOUSE J A and METCALF J A 'The Mechanical Efficiency of I.C. Engines' The Motor Industry Research Association (MIRA), Research Report 1958/5, Warwickshire 1958.

(8) BISHOP I N 'Effect of Design Variables on Friction and Economy' Society of Automotive Engineers. SAE paper 64812 A 1964.

(9) ROSENBERG R C 'General Friction Considerations for Engine Design' SAE Paper 821576. Also in SAE publications SP-532 Aspects of Internal Combustion Engine Design pp 59-70 1982.

(10) PARKER D A and ADAMS D R 'Friction Losses in the Reciprocating Internal Combustion Engine' Tribology - Key to the Efficient Engine. MEP publication pp 31-39 1982.

(11) PINKUS O and WILCOCK D 'The Role of Tribology in Energy Conservation' ASLE Lubrication Engineering. Vol 34, 11, 599-610 Nov 1978.

(12) LANG O R 'Reibungsverluste in Verbrennungs-motoren' (Friction losses in combustion engines). Schmiertechnik + Tribologie 29 January 3 1982. Presented at conference 'Limits of Lubrication' London July 1981.

(13) VICKERY P E 'Friction Losses in Automotive Plain Bearings' SAE paper 75002 1975.

(14) CAMERON A 'The Principles of Lubrication' (Longmans or John Wiley and sons). 1966

(15) JONES G J 'Crankshaft Bearings : Oil Film History' 9th Leeds-Lyon Symposium on Tribology - 'Tribology of reciprocating engines' September 1982. (Butterworths 1983).

(16) BOOKER J F, GOENKA P K and van LEEUWEN H J. Appendix B of 'Dynamic Analysis of Rocking Journal Bearings with Multiple Offset Segments' Trans ASME J.Lub.Tech. October 1982 104, 478-480.

(17) BOOKER J F 'A Table of the Journal Bearing Integral' Trans. ASME Journal of Basic Engineering June 1965 p 533-535.

(18) BOOKER J F 'Dynamically Loaded Journal Bearings : Numerical Application of the Mobility Method' ASME Journal of Lubrication Technology, Vol 93 No 1. Jan 1971 p 168, Errata : No 2 Apr 1971 p 315.

(19) GROBUSCHEK F and EDERER U 'A Contribution to the Evaluation of the Expected Performance of Reciprocating Engine Bearings' 12th International Congress on Combustion Engines, CIMAC Tokyo 1977.

(20) SOMEYA T Discussion on GROBUSCHEK and EDERER paper (ref 19) 12th International Congress on Combustion Engines, CIMAC Tokyo 1977.

(21) SPIKES R H and ROBINSON S M 'Engine Bearing Design Up-to-Date' Proc I.Mech.E. Conf 'Tribology - Key to the efficient engine' January 1982 MEP publication pp 1-8.

(22) MARTIN F A, BOOKER J F and LO P M 'Influence of Engine Inertia Forces on Friction in Connecting Rod Big-End Bearings' to be published.

(23) GOENKA P K 'Analytical Curve Fits for Solution Parameters of Dynamically Loaded Bearings' ASME Paper 83-Lub-33, presented at the ASLE/ASME Lubrication Conference, Hartford, Conn, October 1983

(24) SMITH E H 'Temperature Variations in Crankshaft Bearings' Proc 9th Leeds-Lyon Symp on Tribology. Leeds Sept 1982 (Butterworths 1983).

(25) GOENKA P K 'Dynamically Loaded Journal Bearings : Finite Element Method Analysis' ASME Paper 83-Lub-32, presented at the ASLE/ASME Lubrication Conference, Hartford, Conn, October 1983.

(26) GOENKA P K Private communication

(27) MOES H and BOSMA R 'Graphical Construction of Journal Paths for Dynamically Loaded Bearings'. 3rd International Tribology Congress, Warsaw, Poland, The Polish Tribology Council Vol II pp280-296 Sept '81.

(28) RITCHIE G S 'The Prediction of Journal Loci in Dynamically Loaded Internal Combustion Engine Bearings' Wear 1975, 35, 291-297.

(29) RITCHIE G S and COWKING E W, GEC (UK). Private communication.

(30) BRUNI L and REVELLO P L 'Reduction and Losses due to Friction and Weight of the Reciprocating Masses in Internal Combustion Engines' First IAVD (International Association for Vehicle Design) Congress, Geneva February 1984.

(31) MIYACHIKA M 'Analysis of Engine Journal Bearing Behaviour for Friction Reduction'. Reprint of 1980 Autumn Lectures, Society of Automotive Engineers of Japan, p 385.

(32) GOJON R and SERRE Y 'Calculating a Wear Hazard' Industrial Lubrication and Tribology, Sept/Oct 1983 p 164.

(33) BUSHE N A and ZAKHAROV S M 'Principal Trends in Research Aimed at Improving Reliability of Liquid-Friction Bearings' Soviet Journal of Friction and Wear (USSR Academy of Sciences & Belorussian Academy of Sciences) Vol 1, No 1, pp 68-77 1980, Allerton Press Inc, New York. Trenie i Iznos Vol No 1, pp 90-140, 1980

(34) HOPPE J 'Minimum Allowable Film Thickness, Surface Quality and Running-in Effect in Hydrodynamic Journal Bearings' Proc 4th Leeds-Lyon Symposium on Tribology 'Surface Roughness Effects in Lubrication' MEP Publication 1978.

(35) VOGELPOHL G 'Thermal Effects and Elasto-Kinetics in Self-acting Bearing Lubrication' Proc Lubrication and Wear, Editors D Muster and B Sternlicht. McCutchan Publishing Corporation 1965.

(36) BOOKER J F 'Design of Dynamically Loaded Journal Bearings.' Fundamentals of the Design of Fluid Film Bearings, ASME 1979.

(37) ROSS J M and SLAYMAKER R R 'Journal Centre Orbits in Piston Engine Bearings' SAE Trans, paper 690114, 1969, 78, pp 548-573.

C62/85

The assessment of diesel fuel 'lubricity' using the four-ball machine

P S RENOWDEN, BTech
Lucas CAV Limited, Warple Way, London W3

SYNOPSIS

With the impending introduction of diesel fuels of lower quality, as well as alternative fuels, it has become necessary to measure the 'lubricity' of fuels. Lucas CAV have developed a test procedure for the four-ball machine which enables the load carrying capacity of a fuel to be measured. Attempts to develop a similar test procedure for the assessment of equilibrium friction and polishing wear rate were unsuccessful, due mainly to problems with heat generation. Further work is required and a test procedure which may prove to be successful but has yet to be tested is suggested.

1. INTRODUCTION

Rotary fuel injection equipment (F.I.E.) for diesel engines relies upon the fuel to provide lubrication for all of the many components which are in relative motion and are subjected to load, Fig 1. The modes of operation of these components include, rolling, sliding, and oscillation, as well as various combinations of these. It has been observed that nearly all of the conjunctions at some time operate under conditions of boundary lubrication. This being the case, the ability of the fuel to provide adequate lubrication is dependent upon its ability (by reacting with the surfaces) to form a lubricating film which has the capability of maintaining separation of the surfaces, during operation, Ref (1). The ability of a fuel to give good boundary lubrication is termed its lubricity; though it is now widely recognised that lubricity depends also upon the geometry and composition of the surfaces being lubricated, and the conditions of operation.

Currently, conventional diesel fuels are going through a number of changes which could affect their lubricity; there is also increased pressure for the introduction of light alcohol fuels into the diesel engine.

The test procedure described herein has been developed in order that the lubricity of current diesel fuels can be monitored, and the most effective lubricity additive for alcohol fuels can be identified together with the optimum dosage level.

The monitoring of diesel fuel lubricity would facilitate the avoidance of the type of problem that was experienced with aviation kerosene Ref (2). Here a change in the refinery process produced an unexpected loss of fuel lubricity which was not detected until fuel pump failures (scuffing) occurred in service.

Within the term lubricity, there are three main characteristics that are important:-

(A) Load Carrying Capacity (L.C.C)

This is the maximum load or pressure at which the lubricating film formed by the fuel can maintain viable lubrication and thus prevent scuffing taking place. In terms of F.I.E. operation this controls the maximum pumping pressures that can be used, the ability of the equipment to digest foreign particles, accept thermal shock, and survive general ill treatment.

(B) Polishing Wear Rate (P.W.R)

This is the rate at which material is removed from the lubricated surface as a result of it taking part in the chemical reaction which produces the lubricating film. This produces a high polish on the surface. Some of the rotating components within F.I.E. are manufactured with two-three micron diametral clearances to prevent high pressure leakage, and even relatively small amounts of wear can increase this leakage, to unacceptable levels. Thus fuels which give a particularly high rate of polish could seriously reduce the long term life of the pump.

(C) Friction

This is the resistance to relative motion between the surfaces due to the shear resistance of the lubricating film formed between them. This can become important not only in terms of the power consumption of the fuel pump, but also in terms of heat generation, speed of actuation and in particular cases the speed and manner (rolling or sliding), that components move relative to one another.

Having identified the parameters within the term lubricity that needed to be measured, the methods of obtaining such measurements were considered. The main choice was between using an existing standard tribological machine, and the construction of a new purpose built machine.

Having considered the long term aim of creating a 'lubricity standard', and the general acceptance of the measurement procedure, it was decided to pursue the use of an existing test machine. On the basis of its wide availability the Four Ball Machine was chosen.

2. **FOUR BALL MACHINE**

The machine was originally developed by Shell (Holland) during the early 1930's Ref (3), for testing E.P. lubricants, and it has since been used increasingly, with some variations in construction and facilities for a wide variety of work, ranging from routine tests on lubricants to research on boundary lubrication. References (4) - (30) give some idea of the scope and geographical spread of Four Ball Machine work. These are only the items which are present in the Lucas CAV tribology bibliography and the total number of papers published on the subject is certainly very much larger.

The Four Ball Machine is probably the most widely used laboratory machine for tribological purposes: all major oil companies and additive manufacturers have at least one as have many academic institutions and consultancy organisations. Four Ball Machine procedures are also the subject of U.S. and other national specifications.

The machine consists essentially of four $^1/_2$" diameter bearing balls, three of which are clamped together in contact with each other in a stationary 'ball pot', with their centres all in the horizontal plane and equidistant from the rotational axis of the machine. The fourth (upper) ball is mounted in a rotating chuck and is in lubricated contact with all three lower balls which are not permitted to rotate Fig 2. The load is applied along the axis of the machine via the ball pot and against the fourth ball.

Thus in any given test, a sliding load is applied to one area only on each of the lower balls, and to a continuously traversing 'contact band' on the upper ball. Because of the very short contact duration and the much larger area of this 'contact band' compared with the circular contact areas on the lower balls, wear on the upper ball is usually trivial until scuffing occurs. Wear is assessed by measuring the wear scars on the lower balls. Most machines include a facility for measuring the torque transmitted (the friction torque) to the ball pot. At least one machine Ref (20) is claimed to include an effective means for recording wear continuously during testing, as well as an improved friction torque measurement.

3. **LOAD CARRYING CAPACITY** (L.C.C)

3.1 Test Procedure

Initially the test procedure employed was that detailed in ASTM 2783, Ref (31) although some alterations were required, due to the fact that this procedure is intended for E.P. lubricants with far greater load carrying capacity than diesel fuels. These changes included the removal of the weld load test; this was not thought relevant to F.I.E., where the onset of scuffing is usually catastrophic. Also the first test load was reduced from the recommended 80kg to 6kg. Apart from these changes the test procedure of running for 10 seconds under a specified load, and measuring the wear scars on the 3 bottom balls in both the direction of rotation and at right angles to this, remains as in the ASTM procedure. Subsequent tests (with new balls) are conducted with the load increased by a factor of x1.26 (to the nearest kg). This is continued until a sharp increase in friction or the wear scar diameter is noted, (transition points), in order to determine the machine load at which the lubrication fails and scuffing occurs. Due to the limitations of the machine used for these tests the test speed used was 1425 $\pm$20r/min.

After initial tests were conducted which followed the above procedure the balls were critically examined under the optical microscope, and in some cases the scanning electron microscope, and the load at which the first signs of scuffing occurred noted. Comparing these results with the machine load indicated by the friction and wear transition values, it was found that although there was good correlation between the wear transition values and the point at which scuffing first occurred, no such correlation existed with the values indicated by the friction transition.

In order to detect the wear transition point it is necessary to measure the wear scars on the balls. Having already made this measurement it was found that, rather than quoting the machine load at the wear transition point, a much more accurate and reproducible assessment of the load carrying capacity of a fuel can be obtained by determining the maximum surface pressure under which viable lubrication has been maintained (prior to scuffing). This can be obtained by calculating the conjunction pressure which exists at the end of each test, using Equation (1).

$$P = \frac{0.52\ L}{d^2} \ \text{kg/mm}^2 \quad ...(1)$$

L = MACHINE LOAD (kg)
d = MEAN WEAR SCAR DIAM (mm)
P = FINAL CONJUNCTION PRESSURE kg/mm^2

The load carrying capacity of a fuel (L.C.C) is simply the maximum value of conjunction pressure that has been obtained during any series of load tests. This has the added advantage of removing any need for a subjective assessment of the wear transition load, which in some cases can be extremely difficult.

An idealised plot of a typical conjunction pressure versus machine load curve is shown in Fig 3. Also shown is the diameter of the contact area that would result simply from Hertzian deformation, together with a corresponding plot of Hertzian pressure. Note that the wear scar diameter should never fall below the Hertzian diameter, similarly the conjunction pressure should never exceed the Hertzian pressure.

As the conjunction pressure is proportional to the inverse square of the wear scar diameter, an accurate measurement of this diameter now becomes critical and an accuracy of $\pm$.001mm has been found necessary. In order to achieve this the balls from more recent tests have been measured using a microscope with a micrometer eyepiece attachment.

It is recommended that four such tests be conducted in order to achieve a mean L.C.C. value.

During the tests reported here care was taken not to handle the balls, chuck assembly or ball cup after the cleaning procedure, to avoid contamination with finger grease which could affect the results.

No special precautions have been taken during these tests to control ambient temperature or humidity.

The majority of the tests have been conducted at Brunel University with a Seta-Shell unit, manufactured by Stanhope - Seta, circa 1968 which operates at a constant speed of 1425 ± 20 r/min, and has a mechanical loading system. The original friction measuring device has been replaced by a modern transducer.

The balls used were similar to the ASTM standard (high polish) and were supplied by RHP Bearings Limited. These are a carbon-chrome steel to SAE 52100.

3.2 Retention/Recovery of Load Carrying Capacity

With some of the fluids that have been tested and in particular cetane and wet ethanol + castor oil even though they exhibit similar maximum conjunction pressures they behave very differently as the machine load is further increased, see Fig 4. After scuffing the wet ethanol + castor oil exhibits excellent retention/recovery of the conjunction pressure. With cetane, once scuffing has taken place, conjunction pressure falls off rapidly with further increases in machine load.

In fuel injection equipment localised scuffing which does not lead to serious wear or seizure, but which is self-limiting and then slowly 'healed' by subsequently running is not uncommon. These events are termed 'transient' (as opposed to 'progressive') failures of lubrication. No intermediate behaviour has been observed: either a scuff is transient and healing follows, or it is progressive and leads rapidly to seizure or catastrophic wear.

The ability of the boundary lubrication film (once having failed) to recover rapidly when the local pressure is reduced by scuffing, is obviously important and the results obtained show that different fuels can behave quite differently in this respect.

In order to quantify this feature it is recommended that the machine load at which the conjunction pressure has dropped to 20% of the maximum previously recorded is measured and the machine load at which the maximum conjunction pressure was obtained subtracted from it, see equation (2), to obtain the retention/recovery coefficient.

$$C_R = L_{20} - L_{MAX} \quad .... \quad (2)$$

where,

C_R = Retention/recovery coefficient (kg)

L_{20} = Machine load at 20% maximum conjunction pressure, (kg)

L_{MAX} = Machine load at maximum conjunction pressure, (kg)

For the two examples shown the retention/recovery coefficient is:-

CETANE .... (43-24) = 19
WET ETHANOL + CASTOR OIL .... (90-24) = 66

4. RESULTS

4.1 Load Carrying Capacity

Table I gives a summary of the results so far obtained using the test method described. All these results are from tests carried out at Brunel University, except those shown in brackets which were obtained by Edwin Cooper Laboratories. The results obtained during the second programme used a sequential test procedure which differed from that currently being recommended, and that used during the third programme of tests. However, it is believed that these results are valid and can be included with those from the third programme of tests for purposes of analysis.

4.1.1 Relevance

One of the first considerations in any exercise of this nature is to examine whether the results obtained are relevant to previous experience with similar fuels. The main reason for choosing alcohol based fuels for these first series of tests was the current interest in their introduction as a diesel fuel substitute. Fortunately we have considerable experience of the performance of these fuels both on internal test machines which use actual F.I.E., and with equipment returned from test centres in various parts of the world.

Fig 5 shows the results of tests that have been carried out on a purpose built 'in-house' cam/roller machine, which reproduces the most sensitive and highly loaded F.I.E. components, the cam ring and rollers, Fig 1. The picture is somewhat complicated by the presence of ignition and corrosion inhibitor additives, but generally speaking, dry ethanol does not cause a breakdown, but wet ethanol (12% H_2O) does. The addition of castor oil to wet ethanol is effective in preventing a lubrication breakdown, and although up to 4% of castor oil is required when other additives are present the indications are that a smaller percentage may be effective when it is the only addition besides the water. The lines drawn give an indication of the boundary between successful operation and failure for the three fluids tested. These same trends have been borne out by tests on complete F.I.E., and by examination of equipment that has been returned from various test centres.

On the 4 ball machine these fuels have produced L.C.C. values which are consistent with all our previous information regarding their ability to provide adequate boundary lubrication. These values are; wet ethanol 64, dry ethanol 111, wet ethanol + 4% castor oil 161, and diesel oil 147, 157, 220 kg/mm^2.

More recently seven fuels, all derived from coal, have been tested. They cannot be further specified for reasons of confidentiality. One of the fuels which was identified as having the lowest L.C.C. (104kg/mm^2), was the only fuel which caused an equipment failure on subsequent testing.

In conclusion it can be said that all the results obtained to date are consistent with our previous experience with actual fuel injection equipment.

4.1.2 Repeatability

The results given in Table I show that the repeatability within a test series is acceptable, with the mean percentage variation of each result about the mean L.C.C. for that series being 5% with a standard deviation of 4%. It is believed that these accuracies could be substantially improved upon as the accuracy with which some of the wear scars were measured during the early stages of testing was in some doubt. Later tests carried out by Edwin Cooper Ltd., who are experienced users of the 4-ball machine produced exceptionally good results with a mean percentage variation of the results about the mean L.C.C. value of 1.0% and a standard deviation of 0.5%. These results were obtained on a modern Cameron - Plint hydraulically loaded machine. It is not clear how much the excellent condition of this machine contributed to the accuracy of the results, but at present it is not considered to be a major factor. Far more important is the accuracy with which the wear scars are measured.

With regard to machine to machine scatter, the only fluid that has been tested in both laboratories is dry ethanol. In both cases ethanol of laboratory reagent quality was used, but not from the same source. The two machines used could almost be regarded as being at either end of the range available; at Brunel University the machine being a 16 year old Stanhope - Seta, mechanically loaded version, whereas that at Edwin Cooper is a modern Cameron Plint, hydraulically loaded machine. With different testers, in different laboratories, and with no control over ambient temperature or humidity, the mean results obtained were 106 kg/mm^2 (Brunel) and 119 kg/mm^2 (Edwin Cooper). The standard deviation of these results was 8 and 2 kg/mm^2 respectively. This is an encouraging result.

4.1.3 Reference L.C.C. Values

The only information regarding the fuel L.C.C. required to provide satisfactory operation of F.I.E. in the 'field' is that the fuel currently being supplied appears to be adequate. Therefore to arrive at a realistic figure for the L.C.C. required by F.I.E., sufficient 'on the road' samples to enable a statistical analysis to be made would need to be tested.

So far only three samples of DERV have been tested, giving mean L.C.C. values of 147, 157, and 220 kg/mm^2. This wide range of results from only three samples may be indicative of the major change in L.C.C. that can occur, depending upon the source of crude, and the particular refinery process employed.

With regard to a reference fluid which could be used to calibrate various machines the cetane (n-hexadecane) sample tested appears to be suitable, as it gives a L.C.C. value of 167 kg/mm^2 which is similar to that of DERV. Unfortunately it is likely that this L.C.C. value has been affected by the impurities still present in this reagent (99%) quality cetane Ref (32). If this is found to be the case then alternatives such as dry ethanol may have to be considered.

4.2 Friction and Polishing Wear Rate

4.2.1 Test Procedure, Results

As both of these are measured under conditions where a lubrication equilibrium has been established, they require similar test procedures.

Fig 6 shows the result of the first attempt to conduct such a test. The test employs the same basic procedure as the L.C.C. test but with a 6kg load and the test times are 0.17, 5, 10, 20, 40, and 80 minutes. The friction value is taken as that at the end of the test, and the wear scars are measured on the bottom balls, in order that the total wear volume can be calculated using Equation (3).

$$Vw = 0.0155 d_s^4 - 1.06 \times 10^{-5} L d_s \quad \text{......(3)*}$$

Where
Vw = Total wear volume (three bottom balls) (mm^3)
d_s = Mean wear scar diameter (mm)
L = Machine load (kg)

*The derivation of this formula can be found in Appendix (I).

All the test results contained in this section are with DERV as the test fluid.

The wear rate shown is for the period by which a particular test exceeds the previous test, that is, the wear volume result from the previous test is subtracted from the current test result and the wear rate calculated, using Equation 4.

$$WR_{(n+1)} = \frac{Vw(n+1) - Vw(n)}{T_{(n+1)} - T_{(n)}} \quad \text{..............(4)}$$

Where

WR = Wear Rate (mm^3/min)
T = Test Duration (min)
n = Test Number

As can be seen, both the friction coefficient and the wear rate increase dramatically with the eighty minute test. Examination of the balls from this test revealed as expected, that a lubrication breakdown had occurred. It was also noted that the volume of fluid present in the ball cup at the end of the test was only half that at the beginning. It was suspected that the cause of this failure was overheating and so the next series of tests consisted of five minutes of running, followed by five minutes of non-operation, but with the load maintained. Also the test fluid temperature was monitored at the end of each test. As can be seen from Fig 7 the sudden rise in friction has been eliminated although the fluid temperature and friction continued to rise throughout the test, and there is an upturn in the wear rate during the forty-eighty minute period. In a normal engineering conjunction one would expect both friction and wear rate to decrease with time.

Fig 8 shows a similar series of results but where the test was composed of one minute of test followed by one minute of standing under load. The test load remained at 6kg. Although the friction peaks and then appears to recover to an equilibrium value, the fluid temperature rises more rapidly than in the previous case and is continuing to rise at the end of the eighty minute test. The wear rate has also started to increase after the sixty minute test.

In order to try and control this temperature rise a slight modification to the machine allowed the load to be reduced to 1kg. The previously used test procedure of five minutes test followed by five minutes standing under load was employed and the results obtained are shown in Fig 9. As can be seen both the temperature and the wear rate appear to have reached an equilibrium after thirty minutes of test (total machine time of sixty minutes), although the friction result appears very erratic. It is believed that this is due to the inadequacy of the friction measuring device at these low loads, rather than any true variation in the conjunction friction.

Table 2 shows the comparison between the wear rate during the twenty-forty minute test period, (1kg test) with the corresponding results from the 6kg test, Fig 7. The wear rate at 1kg has fallen by 57% compared with that at 6kg. The area of the wear scar at the beginning of this period, (after twenty minutes), is 59% smaller in the case of the 1kg test load, which means that the wear rate per unit area of contact is very similar for both the 1kg and 6kg test loads. Thus, even though the conjunction pressure at the start of the test period has changed from $21kg/mm^2$ (6kg test), to $9kg/mm^2$ (1kg test), the lubrication mechanism appears to be very similar for both test loads.

The amount of wear that has taken place during the 1kg test means that to achieve the same volume of wear as that achieved with the 6kg test, the test duration would have to be 2 or 3 times as long.

In order to determine the effective resistance to relative motion between the balls that was being produced by the lubricating film per unit area, the friction force at the conjunction was divided by the total wear scar area (scar area x 3). Fig 10 shows the results of such calculations for the three tests carried out with a 6kg load. The friction force/unit area, which is a measure of the effective viscosity of the lubricating film, diminishes with test time and only in the case of the uninterrupted test does it begin to increase as the test proceeds towards 80 minutes. In both the interrupted tests the trend is continuing downwards. This can be accounted for in a number of ways:-

(A) The reduction in the conjunction pressure which has resulted from the increase in the wear scar has brought about a change in the rheology of the lubricating film being formed.

(B) The increase in temperature which is likely to be far greater at the conjunction than that indicated by oil temperature measurements, has caused a reduction in the viscosity of the lubricating film, (temperature viscosity index).

(C) The reduction in conjunction pressure could directly affect the viscosity of the film (pressure viscosity index).

(D) The reduction in the conjunction pressure would be expected to change the film thickness and hence the resistance to relative motion (friction).

4.2.2 Discussion

From the results obtained it is obvious that an ideal test procedure for the measurement of equilibrium friction and polishing wear rate would involve close temperature control of the test fluid at some value below the first transition temperature (yet to be determined), together with the maintenance of a constant conjunction pressure.

To maintain a constant conjunction pressure the wear scar would need to be monitored continuously in order to allow the load to be adjusted with a suitable closed loop system. As far as the author is aware only one such machine is capable of monitoring the wear rate continuously. This is a standard four-ball machine which has been modified by Cranfield Institute of Technology, Ref (20).

If the above proves to be impracticable then on the basis of the results with DERV, a test procedure which involves running for thirty minutes, followed by a period of standing under load, sufficient to allow the oil temperature to return to within 2^oC of ambient before any further testing, might be employed. This is based on the observation that even with the uninterrupted test, Fig 6 a rapid increase in friction and wear rate did not occur until the test time was in excess of 40 minutes. In conjunction with the above it may also be desirable to choose a test load somewhere between 1 and 6kg, possibly 4kg, although this should be avoided if possible as it does require a slight modification of the loading system.

The reduction of the test speed of 1425 r/min is an obvious means of overcoming the heating problem, but it has not been considered as it would involve major modifications to fixed speed machines.

Whatever test procedure is finally chosen, the repeatability of the results would need to be examined as would the applicability of a test procedure developed for DERV to different fuels, such as alcohols (with or without additives present), where the friction coefficient and hence the heat generation rate may be significantly different.

5. CONCLUSIONS

With regard to the assessment of the load carrying capacity of a fuel, the test procedure described gives results which are both repeatable and appear reproducible on different machines at different test centres. All the results to date are consistent with our experience with alternative and conventional fuels, both from in-house tests and with equipment returned from test centres in various parts of the world. A considerable amount of further work needs to be done in the areas of relevance, repeatability, and machine to machine scatter before this procedure could be recommended for incorporation into fuel standards. It is believed that it can already be used as a useful laboratory test to determine the lubricity additive dosage levels required to impart a particular fuel with adequate load carrying capacity. So little is known about the processes involved in the action of additives that it would not be particularly surprising if anomalies were encountered on occasions. The determination of what L.C.C. current F.I.E. requires would involve the sampling and testing of the U.K. DERV that is currently in use.

With regard to the assessment of equilibrium friction and polishing wear rate we are not yet at the stage where a test procedure can be recommended, although the programme of tests has highlighted the problems and some of the possible methods of overcoming them. During any wear test under constant load the conjunction pressure is constantly falling, and unless a means of constantly assessing the wear scar area without disturbing the test can be found then this may prove to be a serious drawback when comparing fuels which give significantly different wear rates.

6. ACKNOWLEDGEMENTS

The author is grateful to the directors of Lucas CAV Limited for their permission to publish this work.

Much of the work reported, including the development of the load carrying capacity test, was carried out under the direction of Mr G Onion, now retired.

The tests at Brunel University were conducted by Mr K Dutta who made a number of significant contributions as did Dr T S Eyre and other members of his staff.

Mr I R Knowles was responsible for the work carried out at Edwin Cooper Limited, (part of the Ethyl Corporation).

This work formed part of a general investigation on future diesel fuels which was conducted jointly with Perkins Engines Ltd and financially assisted by H M Department of Industry. This assistance is gratefully acknowledged.

7. REFERENCES

(1) ONION, G. Reaction film lubrication, Chartered Mechanical Engineer (I Mech E), July 1979, 26, No 7, 61-64.

(2) VERE, R. A. et al (editors) Lubricity of aviation turbine fuels Report of Mod (PE) Fuel Lubricity Panel, January 1976 (editors)

(3) BOERLAGE G D (Shell, Holland) Four-ball Testing apparatus for extreme pressure lubricants. Engineering (U.K.), July 14th, 1933 , 46-47.

(4) LARSEN, R. G. (Shell, USA) The Study of Lubrication using the four-ball type machine Lubn. Engng. August 1945, 1 35-59

(5) FEIN, R. S. (Texaco, U.S.A.) Measurement of wear volume and interpretation of results with four ball machine. Paper at ASME Confce. March 1959.

(6) FEIN, R. S. (Texaco, U.S.A.) Transition temperatures with four ball machine. Paper at ASME Confe. New York, October 1959.

(7) FENG, I. M. and CHALK, H. (Ethyl Corp., S.A.), Effect of gases and liquids in the lubricating fluids on lubrication and surface damage. Wear, 1961, 4 257-268.

(8) KLAUS, E E and BIEBER H E (Penn State U., U.S.A.) Effect of some physical and chemical properties of lubricants on boundary lubrication. ASLE Trans., 1964, 7, 1-10.

(9) ASKWITH, T. C. et al (J Lucas Co. Ltd. Imperial Coll, U.K.) Chain length of additives in relation to the lubricants in thin film and boundary lubrication. Proc. Roy, Soc., 1966, A291, 501-519.

(10) APPELDOORN, J. K. and TAO, F. F. (Esso U.S.A.) The Lubricity Characteristics of heavy aromatics. Wear, 1968, 12, 117-130.

(11) RUDSTON, S. G. and WHITBY, R. D. (BP U.K.) The Effect of model lubricity oil constituents on the wear of steel. Proc. Inst. Petroleum, July 1971, 57, No 556, 189-203.

(12) BARTON, D. B. (Mobil etc, U.S.A) Preferential Adsorption in the Lubrication process of zinc dialkyldithiophosphate. Paper at ASME/ASLE Confce., New York. October 1972

(13) DORINSON, A. (Atlantic Richfield Co, U.S.A.) The additive action of some organic chlorides and sulfides in the four-ball luibricant test. ASLE 72LC-3C-2 1972

(14) LASLAVSKY, Y. S. et al (Oil Refining Research Inst., USSR) Anti-wear, extreme pressure, and anti-friction action of additives. Wear, 1972, 20, 287-297.

(15) BANIAK, E. A. and FEIN, R. S. (Texaco, S.A.) Precision of four-ball and Timken tests and their relation to service performance. NLGI Spokesman, January 1973, 382-386

(16) SANIN, P. I. (Moscow Academy of Sciences) Investigation of anti-wear additives under various loads and at different sliding speeds. ASLE Trans., 1973, 16, No 3, 185-188.

(17) GROENHOF, E. D. and QUAAL, G. J. (Corning U.S.A.) Response of boundary lubricant additives in phenylmethyl silicone. ASLE Jl., 1974, 30, No 8, 389-393

(18) KAWAMURA, M. et al (Toyota, Japan) Electrical observations of surfaces being lubricated. JSLE/ASLE Confce., Tokyo, 1975

(19) KLAUS, E. E. (Penn State U. etc, U.S.A.) Some chemical reactions in boundary lubrication. JSLE / ASLE Confce., Tokyo, 1975

(20) REASON, B. R. (Cranfield Inst., U.K.) The Cranfield four ball machine: a new development in lubricant testing. 3rd Intl Tribology Confce., Paisley U.K., September 1975

(21) TOURRET, R. and WRIGHT, E. P. (Caltex. Lubrizol U.K.) Assessment and development of testing methods by the IP Mechanical Tests of Lubricants Subcommittee. Wear, 1976, 38, 141-151

(22) TOMARU, M. et al (Tokyo Inst. Technology) Effects of oxygen on the load carrying action of some additives. Wear, 1977, 41, 117-140.

(23) TOMARU, M. et al (Tokyo Inst. Technology) Effects of some chemical factors on film failure under EP conditions. Wear, 1977, 41, 141-155.

(24) CZICHOS, H. (Berlin Inst. for Materials Testing, FRG) Influence of asperity contact conditions on the failure of elastohydrodynamic contacts. Wear, 1977, 41, 1-14.

(25) McCARROLL, J. J. et al (BP, U.K.) The reactions at steel surfaces of chlorine and sulphur containing lubricants. Paper C28 at I. Mech E. Tribology Confce., Swansea, April 1978

(26) HSU, S. M. and KLAUS, E. E. (Amoco, Penn State U., U.S.A.) Some Chemcical effects in boundary lubrication, Part 1 Base oil-metal interaction. ASLE Trans, 1979, 22, No 2, 135-145.

(27) KAGAMI, M. et al(Nippon Mining etc Japan) Wear behaviour and chemical friction modification in binary additives system under boundary lubrication conditions. ASME/ASLE Confce., San Fransisco, August 1980

(28) YAMAMOTO, Y. and HIRANO, F. (Kyushu U., Japan) Effects of different phosphate esters on frictional characteristics. Tribology International 165-169, August 1980

(29) SCHUMACHER, R. et al (CIBA - Geigy, Switzerland) Auger electron spectroscopy study on reaction layers formed under Reichert wear test conditions in the presence of extreme pressure additives. Tribology International, December 1980, 311-317,

(30) DORINSON, A. The Significance of Load and Rubbing Time in the Four Ball Lubricant Test. Wear, 1984, 94, 71-88.

(31) ANSI/ASTM D2783 - 71 (re-approved 1976) 9pp

(32) STINTON, H. C. et al A Study of Friction Polymer Formation. ASLE Transactions, 25,3, 355-360.

(33) FEIN, R. S. and KRUEZ, K. L. Chemistry of Boundary Lubrication of Steel by Hydrocarbons. ASLE Tranactions, 1965, 8, 29-38.

(34) BECK, O. et al. On the Mechanism of Boundary Lubrication, Wear Prevention by Addition Agents. Proc Roy Soc, 1940, A177

APPENDIX (I)

The equation;

$$Vw = 0.0155\ d_s^4 - 1.06 \times 10^{-5}\ Ld_s \quad \text{............(3)}$$

where

Vw = total wear volume (three bottom balls) (mm^3)
d_s = mean wear scar diameter (mm)
L = machine load (kg)

applies when the balls used have a radius of 6.35mm and has been derived using the technique suggested by Fein Ref 5, detailed below.

Fig 11 shows the worn area of one of the bottom balls (shaded area), which can be considered as being composed of two flat bottomed circular lenses, both with a base diameter of D (wear scar diameter), and with radii of curvature of r_1 and r_s respectively. r_1 is the normal radius of curvature of the balls and therefore allows the volume of one of the lenses to be calculated.

The calculation of the volume of the second lens, involves the determination of r_s. Considering again Fig 11, if the top (rotating) and bottom (stationary) balls were completely inelastic then the radius of curvature of the wear scar would be r_1. However if we consider the elastic Hertzian deformation of the top ball, that takes place under load, then the wear scar radius becomes r_2. When the load is released then there will be an elastic recovery of the bottom ball which will further modify the wear scar radius from r_2 to r_s. This can also be calculated using standard equations for elastic deformation. Therefore having obtained a value for r_s the volume of the second lens can be calculated and added to that of the first to obtain the total wear volume.

Formula (3) is applicable whether the ball wear scar is concave or convex.

Table 1

FUEL	MAXIMUM P Values (Kg/mm^2)		
	2nd Programme	3rd Programme	Overall Mean
Dry Ethanol	96, 113, 115, 116	96, 102	106
(Dry Ethanol)		117, 121, 117, 120	119
Wet Ethanol (12% water VV)	57, 59, 57	69, 76	64
Wet Ethanol + Castor Oil (4% Medicinal)	171, 171	153, 148	161
Cetane (n-hexadecane)	171, 185	160, 153	167
Derv A	Not tested	148, 142, 156, 143	147
(Derv B)		219, 222, 219 219	220
Derv C		153, 166, 159, 153	157
Methanol (dry)		139, 143, 166, 147	149
Methanol + 1% Castor Oil		164, 148, 160, 162	159
Methanol + 2% Castor Oil		181, 157, 174, 179	173
Methanol + 3% Castor Oil		181, 193, 187	187

Table 2

	WEAR SCAR AREA (mm^2)		CONJUNCTION PRESSURE (Kg/mm^2)		WEAR RATE ($10^{-6}mm^3$/MIN)	
MACHINE LOAD	1Kg	6Kg	1Kg	6Kg	1Kg	6Kg
AFTER 20 MIN.	0.047	0.113	9	21	2.9	6.7
AFTER 40 MIN.	0.068	0.132	6	19		

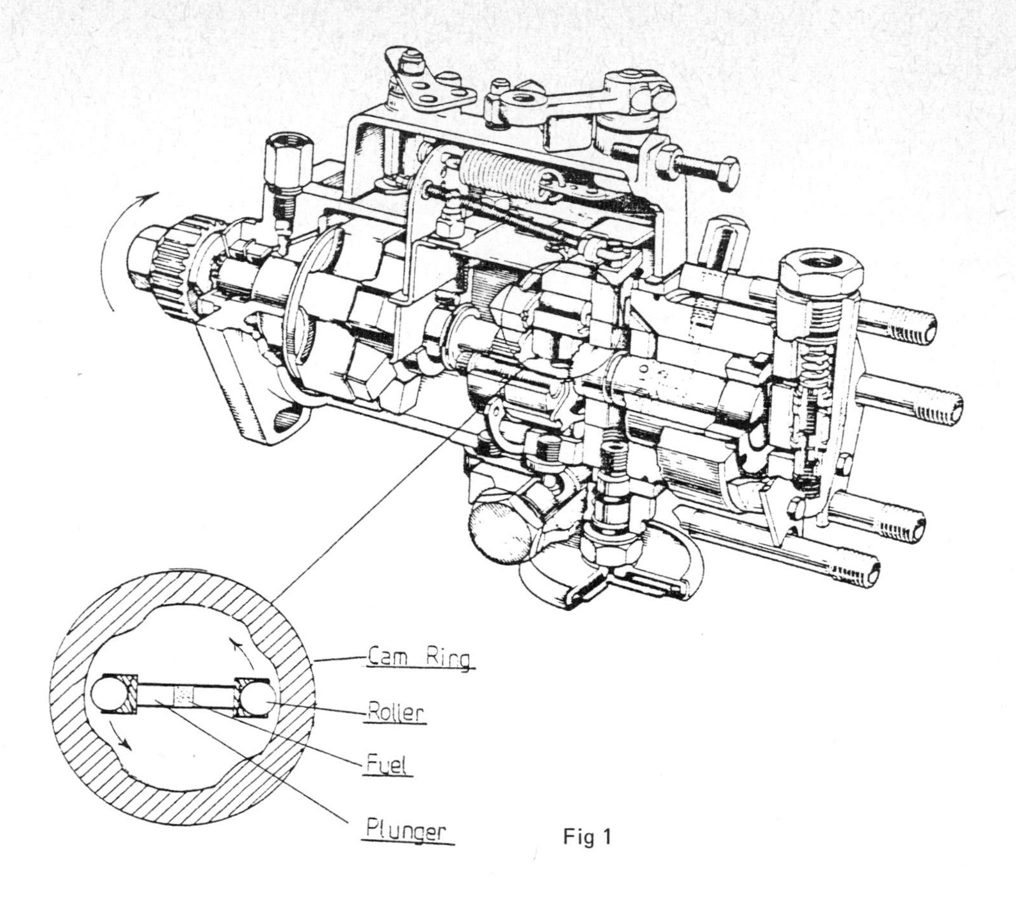

Fig 1

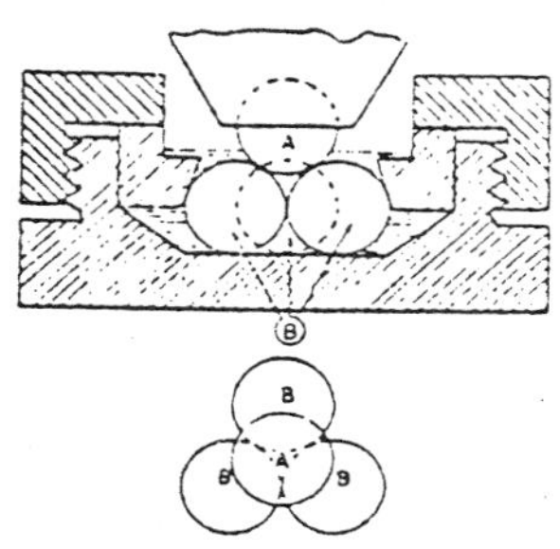

Four Ball Bearing

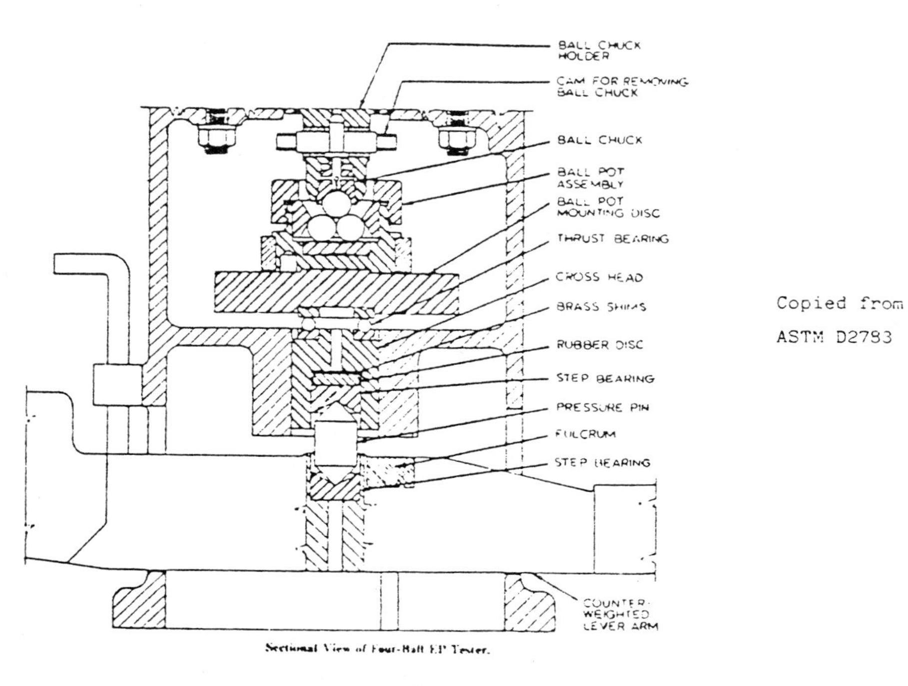

Sectional View of Four-Ball EP Tester.

Copied from ASTM D2783

Fig 2

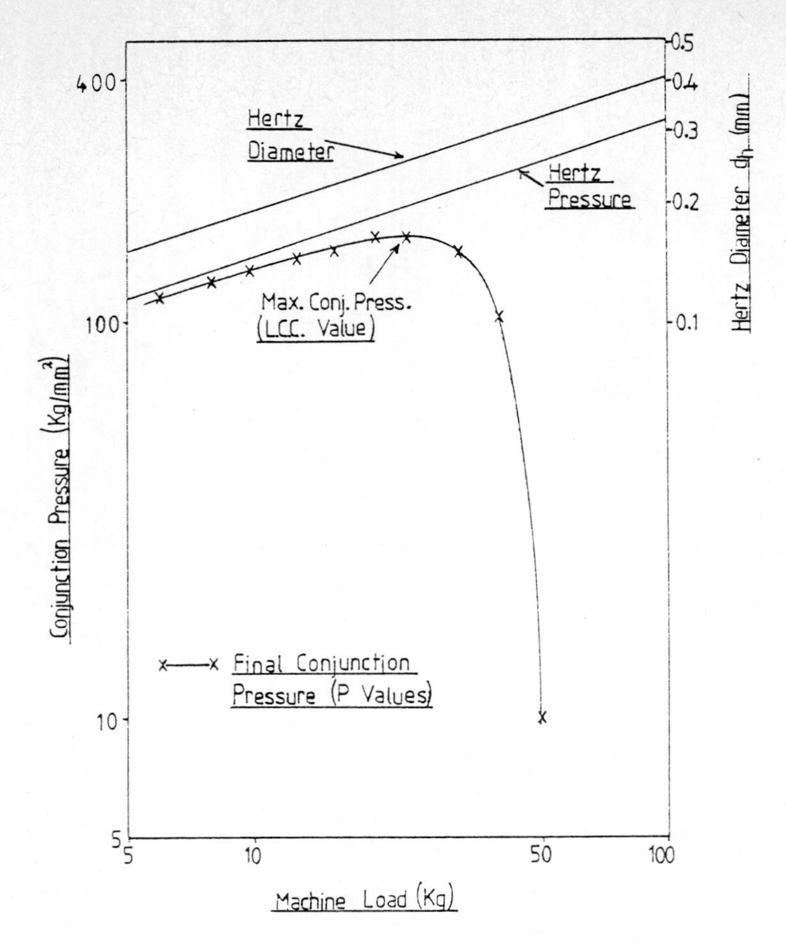

Hertz
Diameter
Hertz
Pressure
Max. Conj. Press.
(L.C.C. Value)
x——x Final Conjunction
Pressure (P Values)
Conjunction Pressure (Kg/mm²)
Hertz Diameter d_h (mm)
Machine Load (Kg)
400
100
10
5
0.5
0.4
0.3
0.2
0.1
5
10
50
100

Fig 3

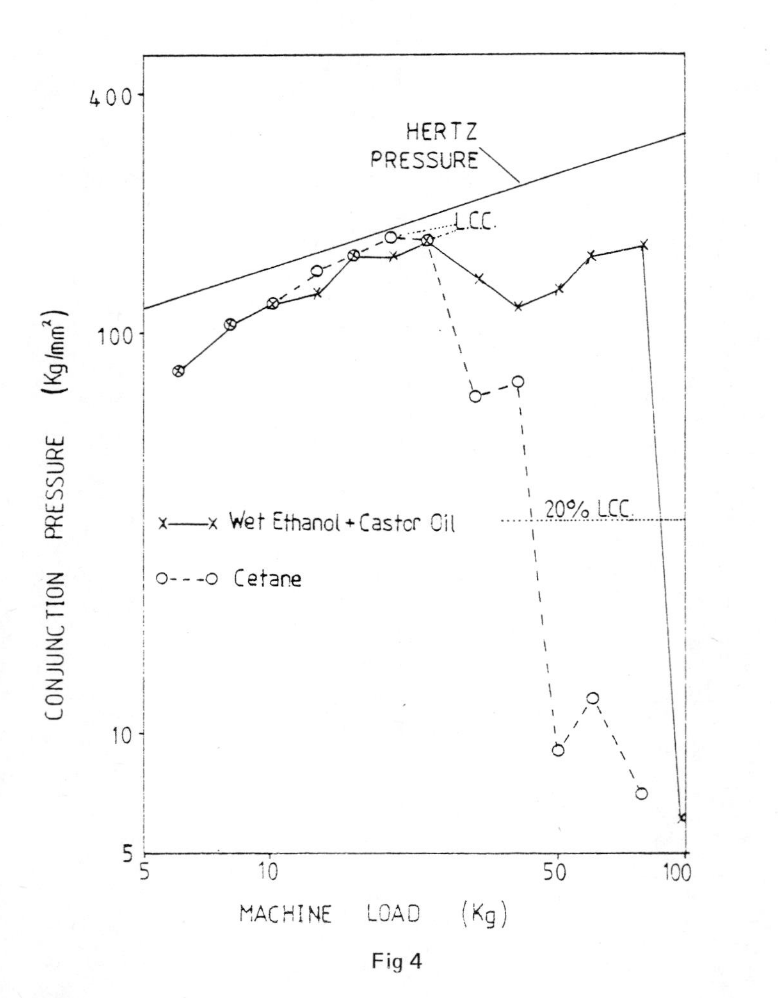

HERTZ
PRESSURE
L.C.C.
20% LCC
x——x Wet Ethanol + Castor Oil
o- - -o Cetane
CONJUNCTION PRESSURE (Kg/mm²)
MACHINE LOAD (Kg)
400
100
10
5
5
10
50
100

Fig 4

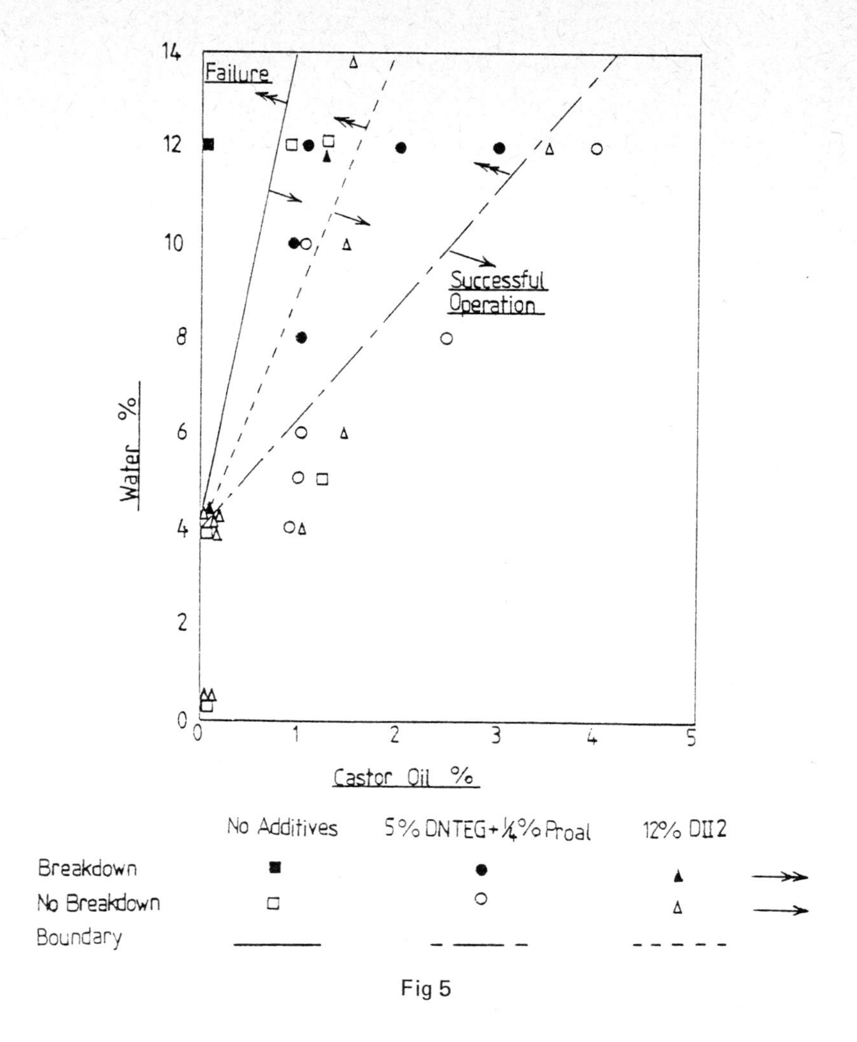
Failure
Successful
Operation
Water %
Castor Oil %
No Additives
5% DNTEG+¼% Proal
12% DII2
Breakdown
No Breakdown
Boundary

Fig 5

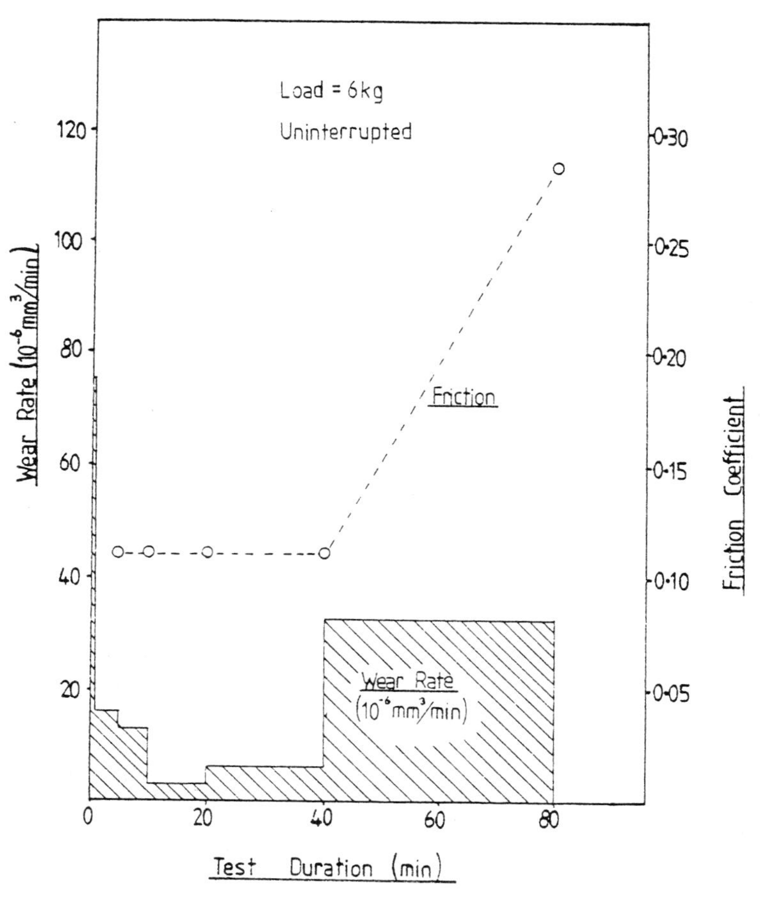
Load = 6kg
Uninterrupted
Friction
Wear Rate
(10^{-6} mm³/min)
Wear Rate (10^{-6} mm³/min)
Friction Coefficient
Test Duration (min)

Fig 6

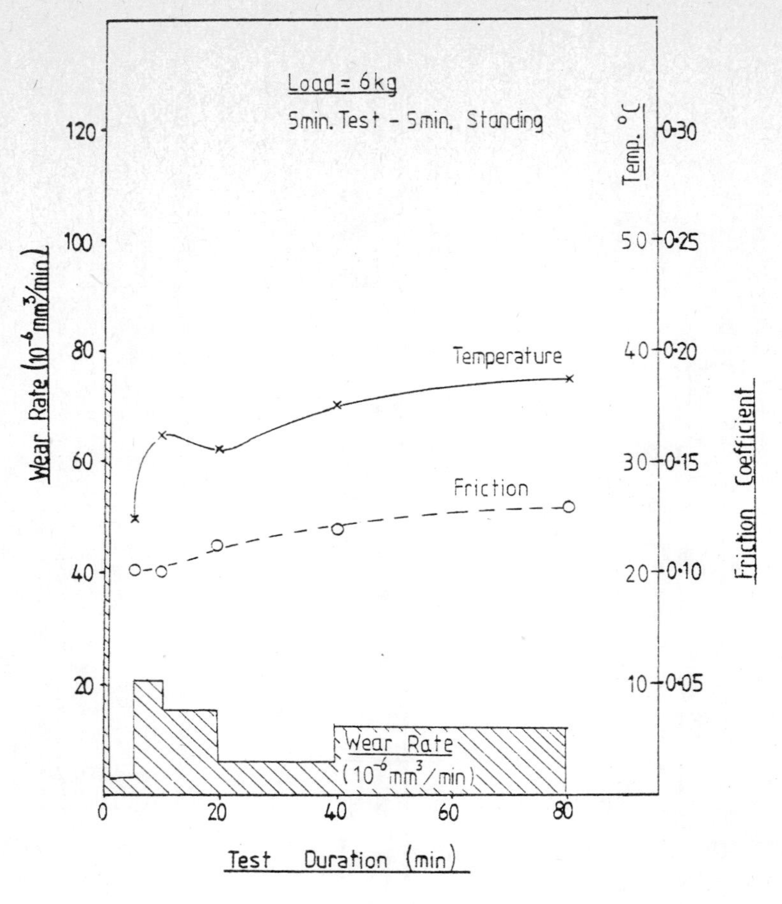
Load = 6kg
5min. Test - 5min. Standing
Temperature
Friction
Wear Rate
(10⁻⁶ mm³/min)
Wear Rate (10⁻⁶ mm³/min)
Temp. °C
Friction Coefficient
Test Duration (min)

Fig 7

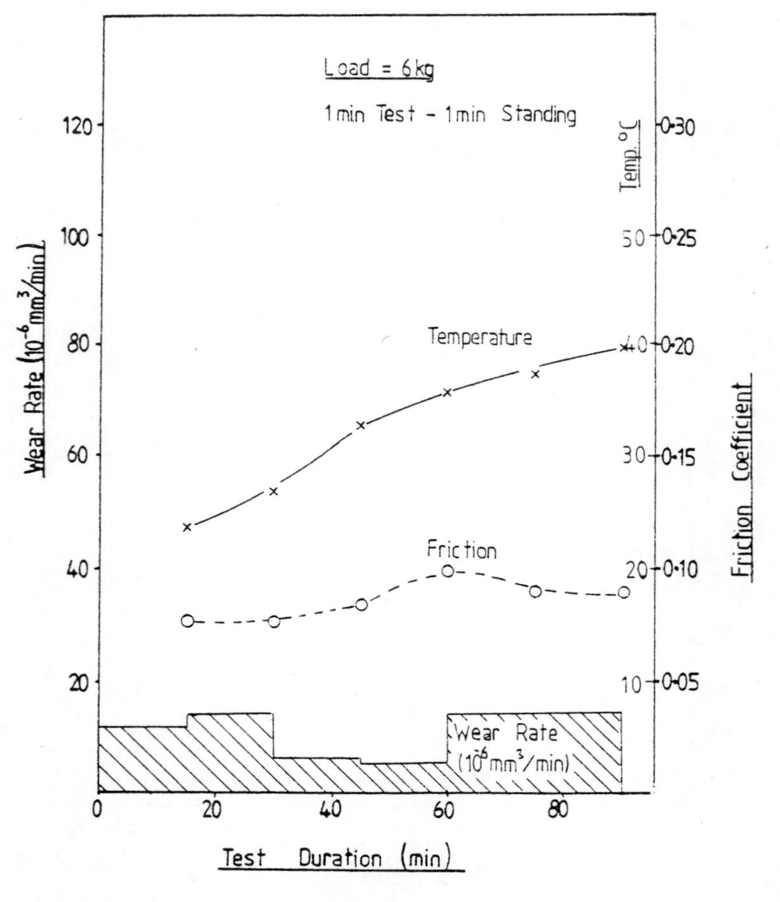
Load = 6kg
1min Test - 1min Standing
Temperature
Friction
Wear Rate
(10⁻⁶ mm³/min)
Wear Rate (10⁻⁶ mm³/min)
Temp. °C
Friction Coefficient
Test Duration (min)

Fig 8

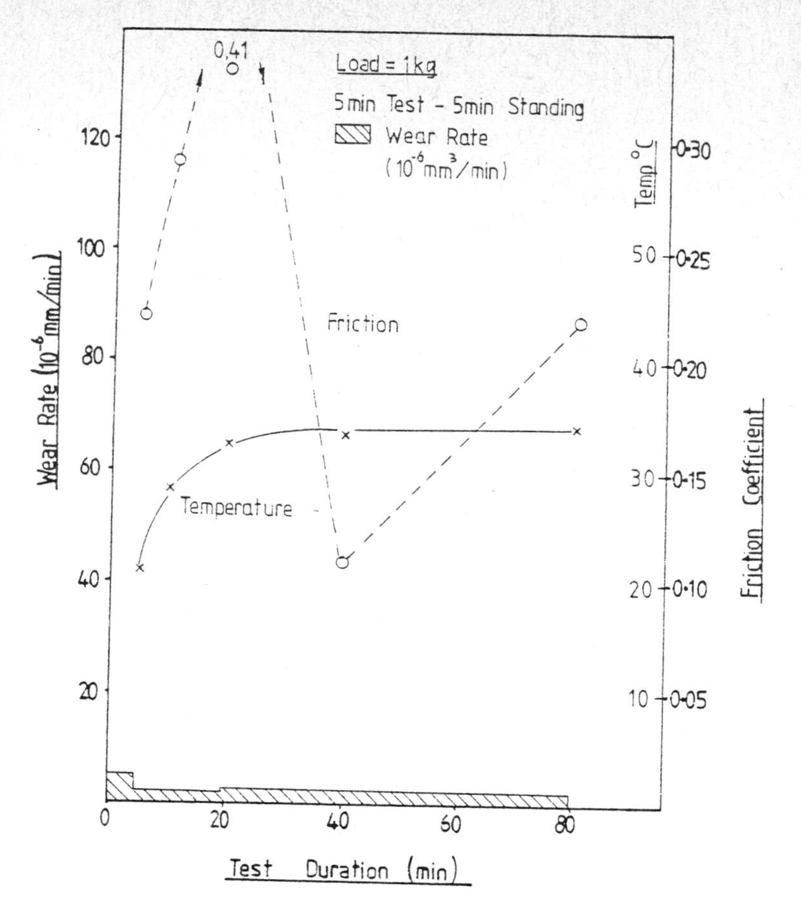
0.41
Load = 1kg
5min Test - 5min Standing
Wear Rate
(10^{-6} mm^3/min)
Friction
Temperature
Wear Rate (10^{-6} mm/min)
Temp °C
Friction Coefficient
Test Duration (min)

Fig 9

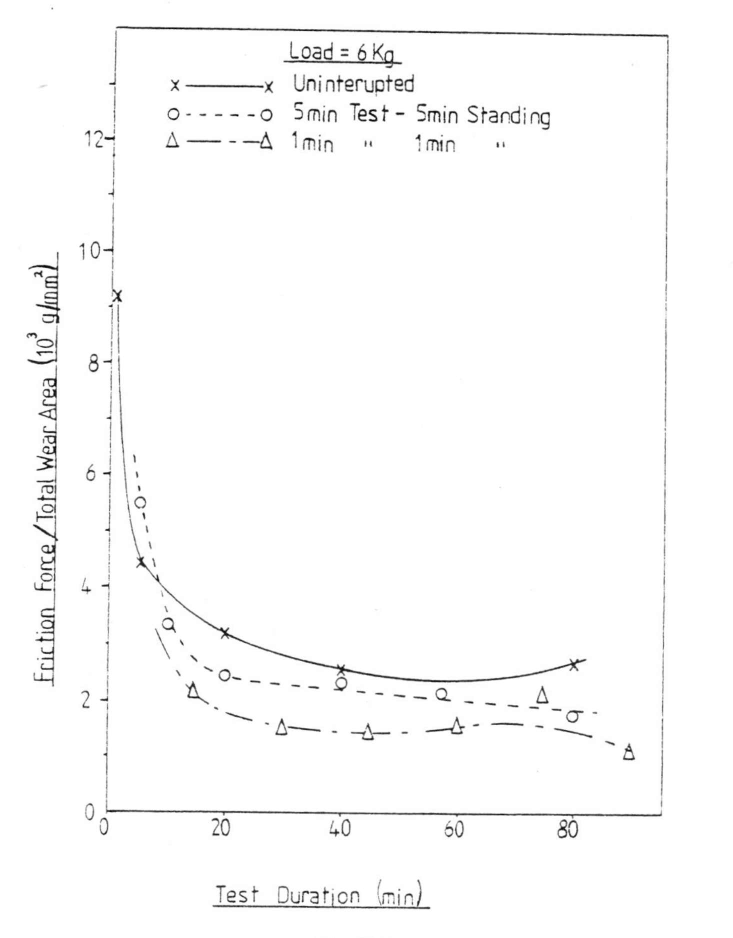
Load = 6 Kg
Uninterupted
5min Test - 5min Standing
1min " 1min "
Friction Force / Total Wear Area (10^3 g/mm^2)
Test Duration (min)

Fig 10

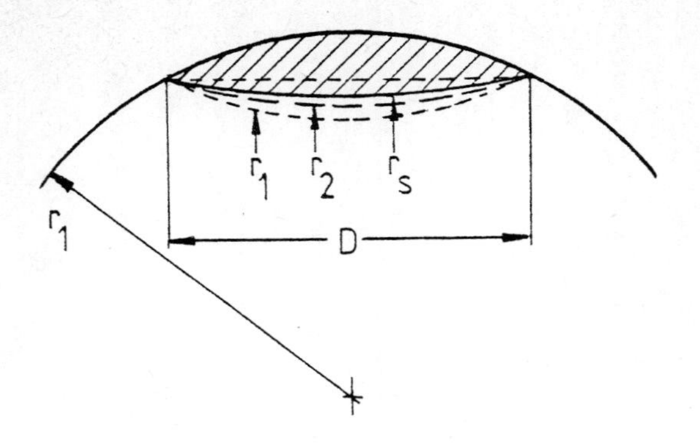

r_1 = Radius of Ball

r_2 = Radius of wear scar (loaded state)

r_s = Radius of wear scar (unloaded state)

▨ = Wear Volume

Fig 11

Possible reduction in fuel consumption by modifying the piston rings

P F KUHLMANN, Dr-Ing
Hochschule der Bundeswehr, Hamburg, West Germany
R J JAKOBS, Dr-Ing
Goetze AG, Burscheid, West Germany

SYNOPSIS The paper presents an investigation which is to show what reduction in fuel consumption in spark ignition engines is actually possible by modifying only the piston rings.
This investigation includes a mathematical estimation and engine tests with highly precise fuel consumption measurements.

1 INTRODUCTION

The share of the piston assembly in the overall friction of an engine (difference between indicated power and brake power) is about 40 %.
It is therefore understandable that efforts are being made worldwide to reduce the friction of this group of components, in order to reduce the fuel consumption of the engine.

In this connection, the extent to which it is possible to reduce friction and fuel consumption by modifying only the piston rings has been estimated. This estimation includes a mathematical theoretical part and engine tests with a passenger car gasoline engine.

Commissioned by Goetze AG these engine tests have been carried out by the Laboratory for Internal Combustion Engines at the Bundeswehr-Hochschule Hamburg.

2 MATHEMATICAL ESTIMATION

2.1 Piston ring friction

To estimate mathematically the friction of the rings, a computer program developed at Goetze AG is used for calculating the oil film between ring and cylinder (1).

The oil film is calculated by means of the one-dimensional Reynolds equation, Fig. 1 left. Apart from the hydrodynamic bearing capacity, that is, the wedge effect, this equation also takes into account the squeeze effect at the dead centres.

The following assumptions are involved in oil film calculation according to Reynolds equation:

- laminar flow
- completely filled lubricating gap
- inertia force negligible compared to friction force
- incompressible Newtonian fluid as lubricant
- constant viscosity
- possibility of radial movement only, exactly perpendicular to the cylinder wall, by the piston rings within the groove, friction on the sides of the ring not considered
- no oil flow in circumferential direction
- no influence of secondary movement of the piston
- O. D. profile of the piston ring assumed to be an arc of a circle
- entire length of the liner is a circular cylinder

The Reynolds equation provides the pressure distribution along the width of the ring. The radial gap between ring and cylinder results from the balance between the oil film pressure, the ring tension and the gas forces on the back of the ring.
The friction force per unit of circumference of the ring then follows from the Newtonian shearing stress hypothesis, Fig. 1 left. The friction forces represented in the following diagrams are summed up over one engine cycle.

The cross-sectional profile of slotted oil control rings, with or without spring, cannot be calculated by means of the existing computer program. This estimate is therefore made in a greatly simplified form by the two contact lands of the oil ring each being calculated individually as plain compression rings (each with half the tangential force).

The barrelling of the profile occuring in practice with these oil ring types is almost zero. However, since according to the Reynolds equation no pressure build-up is possible for plane-parallel surfaces, the peripheral profile of the oil ring lands is also assumed to be an arc of a circle, but with a very shallow curvature.

The effect of the various piston ring variables on the piston ring friction is estimated mathematically for a gasoline engine with the data:

Bore 85 mm
Stroke 86 mm

for the two operating levels

n = 2100 rev/min lower partial load
n = 5000 rev/min full load.

Fig. 1 right, shows the applied pressure curves above and below the top compression ring. The pressure curves above the top compression ring are measured gasoline engine combustion pressure curves, while the pressure curves below the top ring were assumed to be analogous to pressure relationships according to Furuhama (2). No differential pressure between top and bottom sides was assumed to exist for the oil ring.

The O. D. profile of the compression ring is specified as an arc of a circle - indicated by the peripheral radius R. The O. D. radius is then kept constant for each of the implemented variations in ring width ($R = 42.5$ mm $\hat{=}$ $D/2$, D = ring diameter). The O. D. profile of the oil ring lands is also assumed to be an arc of a circle that is kept constant for all land width variations. Here, O. D. radius of $R = 42.5$ mm $\hat{=}$ $D/2$ is used, too.

The viscosity of the lubricating oil for the oil film computation is assumed as a constant (10 cP). The extent of the influence of different viscosities under the ring parameters named was not studied.

Fig. 2 shows as an example the result from the calculations for the top compression ring. Shown is the effect of ring width and tangential force or surface pressure on ring friction at full load.

According to this figure, friction decreases as ring width is reduced. A close evaluation of the curves reveals that, beginning at ring width over 1.5 mm, a reduction in ring width at constant surface pressure p_o has a stronger effect on friction than a reduction in tangential force to the same value but at constant ring width. For initial ring widths of 1.5 mm or less, however, both measures result in approximately the same amount of friction reduction.

Fig. 3 shows friction losses per contact land of an oil ring, as related to the land width for two different surface pressures at full load and partial load. The friction continuously decreases as land width is decreased.

2.2 Assumptions for the estimation of the fuel consumption reduction

Apart from the effect of the various piston ring variables on the ring friction, however, the saving expected in fuel consumption is also to be estimated.

For that it es assumed that the piston rings account for 20 % of the friction mean effective pressure p_r. The friction losses p_r are defined as the total difference between indicated mean effective pressure p_i and brake mean effective pressure p_e.

It is further assumed that the oil ring accounts for 60 % of the friction of the complete ring pack, with the compression rings of the three-ring piston each taking up 20 %.

Two different assumptions can be made when deriving the relationship between the specified reduction in friction mean effective pressure and the resultant saving in fuel consumption.

A: The reduction in friction mean effective pressure is converted at the same indicated horsepower to a higher brake horsepower (for example, by changing the gear ratio).

B: For the same brake horsepower, a reduction in friction mean effective pressure causes a lower indicated horsepower.

As a first step, assumption B will certainly be closer to actual practice and is applied for the sample calculations.

Fig. 4 shows the relationship

$$\frac{\Delta b_e}{b_e} = f\left(\frac{\Delta p_r}{p_r}, \eta_{m_o}\right) \text{ for } p_e = \text{const. } (b_e \hat{=} \text{BSFC})$$

as a formula and as a graphic representation for the range of interest.

The mechanical efficiency η_m of gasoline engines decreases to about 0.45 at the lowest range of partial load, whereas values of about 0.80 occur at the rated power point (3). The following sample calculations have therefore been based on η_{m_o} = 0.45 and 0.80.

2.3 Results

The possible savings in fuel consumption are estimated mathematically as follows for certain friction-reducing modifications at the compression and oil rings, Table 1.

The following modifications have for example been calculated (without taking into account a possible impairment in the functional characteristic, that is, oil consumption and blowby):

1 Reduction of compression ring width from 2.0 mm to 1.5 mm at p_o = constant.

2 Further reduction of compression ring width from 1.5 mm to 1.2 mm at p_o = constant.

3 Reduction of the tangential force F_t of a 1.5 mm width ring to the value for a 1.2 mm width ring.

4 Reduction of the F_t of a 1.5 mm width ring by 50 %.

5 Reduction of the land width of an oil ring from 0.40 mm to 0.25 mm at p_o = constant.

6 Reduction of land width according to modification 5 as well as an additional reduction of surface pressure from 1.34 to 0.6 N/mm².

These sample calculations indicate that:

- The possible reduction in fuel consumption for a compression ring is, assuming current production ring design, very slight, being 0.14 % to 0.25 % at low partial load for modifications 1 - 3 .

- Even assuming an F_t reduction of 50 % on the compression ring (modification 4), which can hardly be implemented in practice, only a maximum of 0.35 % reduction in fuel consumption can be realized at the lower range of partial load operation.

- The possible reduction in friction and fuel consumption through a change from a 1.5 mm width ring to a 1.2 mm width ring (at p_o = const.) can also be achieved to the same degree by reducing the F_t on the 1.5 mm ring. Possible increase of wear on the O. D. and sides, which can occur sooner or later to rings smaller than 1.5 mm depending on the engine, is thereby avoided.

- The overal reduction in fuel consumption that is possible for the oil ring is greater, and can be as much as 1.8 % in the low range of partial load at a drastic reduction of surface pressure.

- The possible reduction in fuel consumption for full load amounts to only one third of the reduction achieved for the low range of partial load.

3 MEASURING METHODS AND MEASURING ACCURACY

The mathematical estimates mentioned above showed that by design modifications to the piston rings only slight changes of the fuel consumption are to be expected. This means that the highest effort has to be made in the experimental determination of specific fuel consumption. The concern here is not merely to achieve the highest accuracy in the measured values but also to exclude side effects which might influence the fuel consumption and mask the effect of the modified piston rings. In total the following points are to be observed:

1) Accuracy of the measured values;
2) Accuracy of setting the operating level of the engine;
3) Feasibility of reproducing the same test conditions;
4) Influence of uneven running of the engine;
5) Variations in the condition of the engine.

In the following text all these points will be examined more closely, and the steps taken to increase measuring accuracy and to eliminate inaccuracies will be described.

3.1 Accuracy of the measuring instruments

To determine the power, the speed and the torque are measured. The inaccuracy of the speed measurement only amounts to ± 1 rev/min and can therefore be ignored. The torque is measured via a load cell (resistance strain gauge type) on the dynamometer brake and a reading exact at 0.01 Nm can be taken. Numerous calibrations of the torque measuring equipment in the course of one year showed a deviation of the calibration factor of only 0.15 %. However, it became obvious that the zero position had shifted in the course of that period by 1 Nm, which would mean an unacceptably serious inaccuracy. For this reason, the zero position and the calibrating factor has been checked before every important measurement.

To determine the fuel consumption, two volumetric measuring instruments are connected in series (flow-meters from Pierburg and from Seppeler). In the course of numerous measurements the deviation of the readings of the two measuring instruments only amounted to 0.1 %, from which it can be concluded that each of them achieves at least this degree of accuracy. The density of the fuel is calculated according to temperature, where the value at 15 °C for a tank filling of 3000 liters is repeatedly determined with the aid of an areometer. An inaccuracy in fuel density at 15 °C does in fact decrease the absolute accuracy of the measured values, which however, does not concern the comparison of one measurement with the other, which is in any case the decisive factor in determining the influence of the piston rings.

To summarize, it is possible to achieve purely metrologically a reproducibility of the measured values of the specific fuel consumption of about 0.2 %. In practice, however, the reproducibility is not so good on account of fluctuations in the running of the engine, as will be demonstrated.

3.2 Setting the operating level

The speed of the engine is controlled with an accuracy of ± 2 rev/min by the dynamometer brake. But even greater deviations from the set level have no practical effect on the specific fuel consumption, for power and fuel consumption are equally influenced by the speed.

In the matter of the torque setting the situation is different. The smallest alterations to the throttle valve setting and even random fluctuations in the running of the engine will lead to the torque not being exactly adjusted. For this reason the torque is printed out many times during the period of the fuel measurement and eventually the average of the recorded values will be used. Then the fuel consumption at the nominal torque setting can be calculated from the actual fuel consumption measured at the actual torque; here the fuel consumption is assumed to be proportional to the indicated mean effective pressure. At low load in particular this is much more accurate than determining the specific fuel consumption from the measured values of fuel consumption and power.

3.3 Reproducibility of test conditions

It is, of course, very important to know to what extent the results of tests may be reproduced which are carried out on different days or even over longer periods. To improve this reproducibility it is necessary to establish those boundary conditions which can have an influence on the specific fuel consumption. It is then necessary to ensure that they are maintained at a constant level. Alternatively their influence must be quantified so that their influence on the measured fuel consumption can be eliminated by a mathematical correction.

Probably the greatest influence on the specific fuel consumption is the excess air ratio. Accordingly whenever a measurement is taken, the excess air ratio λ is varied via the electronic injection system and the curve BSFC = f (λ) is drawn. The excess air ratio is determined from the exhaust emission analysis according to Spindt (5).

It would be possible to vary the excess air ratio while maintaining a fixed throttle position. This would mean, however, that the indicated power and the brake horse power would decrease with the increasing excess air ratio. A test carried out in this way would not only be unrealistic (at partial load, the power is determined by the driving conditions); it would also affect the reproducibility of the measurements, because the position of the throttle valve would influence the measured values.

Consequently the more reasonable way was chosen, viz. to keep constant the brake horse power while varying the air excess ratio (compare also (6)). Furthermore this means that the mechanical efficiency remains constant and that the minimal values of the specific fuel consumption are to be found at high excess air ratios.

When considering the test results, it is possible to compare the minimal values of the specific fuel consumption. This would seem to have the advantage that inaccuracies made in determining the excess air ratio do not affect the comparison. In fact, however, it has been shown that the reproducibility of measurements of the minimum specific fuel consumption is relatively poor, see Fig. 5. The reason for this is that ignition and combustion are significantly impaired when the excess air ratio is high, and consequently even minute changes in the engine (e. g. in the condition of the spark plugs) will lead to a wide scattering of measured values.

Therefore, it is better to compare the results of measurements made with different piston rings at a smaller excess air ratio, say at λ = 1.1 or 1.2, even though this means the inaccuracy of the excess air ratio measurement is included in the comparison.

Further boundary conditions which may influence the results of the measurements are here:

Cooling water temperature:
Preliminary tests showed that a rise in cooling water temperature of 1 °C reduced friction torque by 0.06 Nm, which led to a reduction of the specific fuel consumption of 0.06 to 0.15 %, depending on the load of the engine. Though the water temperature is nearly kept constant by the thermostat, the measured results are corrected corresponding to the water temperature readings, in order to eliminate even small variations.

Temperature of the lubricating oil:
The influence of the lubricating oil temperature is smaller than that of the cooling water temperature; a rise in temperature of 1 °C behind the oil pump leads to a reduction of friction torque by about 0.02 Nm. Nevertheless, the measured values are also corrected in relation to the oil temperature.
For the correction of the measured values a polynomial function has been worked out which shows the mean friction pressure as a function of speed, cooling water temperature, and lubricating oil temperature. Now the change in the mechanical efficiency can be determined corresponding to the measured mean effective pressure, which leads to the specific fuel consumption of the engine under nominal operating conditions.

Ignition timing point:
Since the throttle valve setting is altered with the variation of the air ratio, the ignition timing point would also alter with the vacuum in the intake manifold. For this reason, when measurements are taken, the vacuum timing advance is suspended, and the ignition timing is set at a fixed point for every operating level.

Atmospheric conditions:
The test engine is located on a test bed with a powerful thermostatically controlled ventilation system. Throughout almost all of the tests, the temperature of the intake air was therefore 21 $\pm$ 1 °C.

3.4 Uneven running of the engine

From the beginning it was evident that the power of the engine fluctuated a little while the fuel consumption measurements were being taken. This became evident by torque fluctuations because the adjustment of the dynamometer brake kept the speed constant. When the adjustment of the dynamometer brake was switched to constant brake torque as a test, then variations in power manifested themselves in a fluctuating speed.

In order to eliminate these variations, steps were taken to keep air feed and fuel feed absolutely constant. The throttle valve was screwed into a fixed position and the control pulses for the electronic fuel injection was not taken from an air flow sensor but from a constant voltage supply.

This produced an improvement evident in the fact that fuel consumption measurements taken one after another now deviated by only 0.05 % from each other. But even now when the speed is adjusted to a constant, there still appear slight variations in the torque of about 0.1 %. These fluctuations are partly the product of cyclic fluctuations of the combustion. This hypothesis is supported by the fact that they increase as the excess air ratio increases. However, it cannot be excluded that this effect is the result of random variations in friction. It must be remembered that a difference in torque of 0.01 Nm at a rate of 2000 rev/min, in theory still readable on the digital instrument, corresponds to a difference in power of only 2 W.

3.5 Condition of the engine

It is most difficult to estimate inaccuracies which may occur in the course of measurement as a result of changes in the condition of the engine. Of particular importance in this context is the running-in of the whole engine, of the pistons and especially of the piston rings. These influences will be discussed in greater detail in the next section in the light of the test results.

However, other changes can also take place. In preliminary tests for example, it was noted that a higher spark plug gap, enlarged by burning off, led to a slight improvement of the fuel consumption, in the case of a high excess air ratio.

Another question is whether the comparison of test measurements may not be impaired by stripping and re-assembling the engine when changing the piston ring packs. However, measurements made before and after removing the cylinder head indicated no differences in consumption levels.

Finally, due attention must be given to the lubricating oil. Corresponding to normal operating conditions, a common brand of multigrade oil is chosen, and a sufficient quantity for all the tests was stored. In order to preclude the influences of the ageing of the oil in the engine, the identical test program from changing the oil to measuring the fuel consumption is also carried out with the comparative tests. It has been shown, however, that the influence of the ageing of the oil is relatively small.

4 TEST AND TEST RESULTS

All tests were carried out on a 4-cylinder spark-ignition engine with electronic fuel injection, the rated power of which is 81 kW at 5400 rev/min. Electronic injection can yield advantages here, since the excess air ratio can easily be varied during running, and since subjecting the electronics to a constant voltage control an extremely uniform injection can be achieved.

All measurements were taken in the course of a pre-established running-in and measuring program. They referred to:

- specific fuel consumption as a function of excess air ratio
- friction losses in the form of torque measurements
- oil consumption at rated speed
- blowby volume at various points of the performance graph.

The BSFC-measurements and the motored torque measurements were taken at four points of the performance graph which correspond approximately to an automobile travelling at:

120 km/h (ca. 75 mph) in 4th gear
90 km/h (ca. 56 mph) in 4th gear
50 km/h (ca. 30 mph) in 4th gear
50 km/h (ca. 30 mph) in 3rd gear

To save space, however, only the results of the operating levels with the largest and smallest loads will be reproduced here:
these are BMEP = 4.32 bar at 120 km/h in 4th gear and BMEP = 1.14 bar at 50 km/h in 3rd gear.

4.1 Variation in tangential force

The purpose of this series of tests was to establish the extent of possible fuel savings when reducing the tangential force of the piston rings. To this end ring packs were chosen in which the rings (compression rings as well as oil rings) lay at the maximum and minimum tolerances of tangential force permissible according to the drawings. In this manner the tangential force of the F_{tmin}-rings was on the average by 28 % lower than that of the F_{tmax}-rings.

First the engine was fitted with rings of maximum tangential force, then with rings of minimal tangential force, and finally the F_{tmax}-rings were tested again. Both types of ring had to undergo an identical running-in program so that they were measured in approximately the same 'run-in' condition.

It is important to note that the torque measurements at motoring made on the F_{tmax}-rings in the first and second series of measurements indicated differences of only 0.1 Nm. From this it may be concluded that the 'run-in' condition of the engine hardly altered during the course of the measurements, and that the differences actually measured between the rings with maximum and minimum tangential force may be attributed to the rings themselves.

The most significant results of the tests habe been compiled in Table 2 and Fig. 6. As anticipated, a smaller tangential force leads to lower friction losses manifest in a reduction of the torque at motoring and of the specific fuel consumption. The absolute reduction of the torque at motoring at operating level "50/3" was 0.9 Nm and at the other operating level 0.7 Nm.

If one derives the specific fuel consumption from the measured reduction of the torque at motoring under the assumption of a constant indicated thermal efficiency, then a BSFC-reduction of 0.7 % at operating level "120/4" is to be expected and a reduction of 2.2 % at operating level "50/3" respectively. Indeed in Fig. 6a the anticipated improvement of 0.7 % can be seen (exactly 0.6 % for $\lambda = 1.1$ and $\lambda = \lambda b_e min$). However, in Fig. 6b, representing the "50/3" operating level, we find instead of the expected 2.2 % improvement one of only 0.6 % for $\lambda = 1.1$ and $\lambda = \lambda b_e min$. From this the significant conclusion may be drawn that it is rather hazardous to estimate fuel consumption from friction losses without direct measurements of the fuel consumption.

Deductions about improvements to fuel consumption using rings of lower tangential force should not be made before taking a look at oil consumption and blowby volume. As Table 2 indicates, with the F_{tmin}-rings oil consumption at rated speed was more than twice as high in comparison to the F_{tmax}-rings. The blowby volume was only slightly influenced by the tangential force of the rings at most points of the performance graph. Only at high speed and no load the increasing blowby volume indicated an instability of the F_{tmin}-rings.

4.2 Variation in ring width

In this series of tests the production ring pack is compared with a ring pack with a lower axial width. To do so the compression rings in the first and second groove have been reduced from 2.0 to 1.5 mm and the land width of the oil control ring has been reduced from 0.4 to 0.25 mm.

The contact pressure is held constant for the variations in ring width.

Moreover the overall width of the oil control ring with the narrower lands has been reduced to h = 3.0 mm (production h = 4.0 mm), in order to keep the conformability of the ring constant in spite of the reduced tangential force.

For these investigations a new engine of the same type was used. Pistons were of course changed with the rings. First a 54-hour running-in and preliminary trial program was carried out using both ring packs one after the other, and subsequently a 28-hour test program was run using one ring pack after the other.

In Table 3 and Fig. 7 the results of both the 28-hour test programs are compared with each other. From the figures it can be seen that the narrow rings indicate a somewhat higher oil consumption, a somewhat lower blowby volume and a fuel consumption lower by 0.6/0.8 % for λ = 1.1 and 1.2/1.9 % for $\lambda = \lambda_{b_e min}$. Surprisingly the measurements of torque at motoring show no difference between the two ring versions.

It is not obvious from the values in Table 3 that the production-size rings exhibited a truly stable character, i. e. oil consumption and blowby volume decreased slightly during running-in the rings. On the contrary, the narrow rings showed strong variations in behavior: oil consumption at rated power increased from firstly 11 g/h to 36 g/h within an operating time of 34 hours. The blowby volume at high speeds and no load had increased considerably in this time as well. These changes over the course of time can be explained by severe wear (especially of the lower side faces) of the top piston rings, which was determined when measuring the rings subsequent to the tests. The axially narrow compression rings in the first groove therefore cannot be used in this type of engine (4).

5 SUMMARY

In order to clarify the basic correlations and to estimate the extent of the fuel savings, calculations on piston ring friction have been carried out. From these investigations it results that - depending on the design modifications to the piston rings - fuel savings of 0.1 to 0.6 % at full load and 0.2 % to 1.8 % at partial load can be determined.

In addition to the theoretical investigations, measurements have been carried out on a spark-ignition car engine. In taking greatest care, a reproducibility of the BSFC-measurements of 0.3 to 0.5 % (depending on engine load) could be achieved. This reproducibility is just sufficient to prove fuel savings by modifications to piston rings.

The measurements showed that by reducing the tangential force of the piston ring pack by 28 %, a reduction in fuel consumption of about 0.6 % - as an average of the two shown operating levels - could be achieved. This reduction in tangential force produced, however, an unacceptable increase in oil consumption.

Measurements that were carried out alternatively with rings having a narrower width showed reductions in fuel consumption of 0.7 % on the average of the two shown operating levels at an excess air ratio of 1.1 and reductions of 1.6 % on the average with minimal specific consumption values. However, the rings in the 1st piston groove exhibited an extremely severe side face wear - already after a short running period which led to a considerable increase in oil consumption.

The measurement results shown are strictly speaking only valid for the engine type tested. But as a result of all those investigations it becomes obvious that with a constant number of piston rings, modifications to the rings alone can hardly effect a significant reduction in fuel consumption with satisfactory performance. This applies especially when the production ring pack, which represents the initial status for the ring pack modifications to be tested, has already been optimized to a great extent.

A more obvious reduction in fuel consumption only becomes conceivable through additional modifications to the piston and possibly to the entire engine power unit, for example, by reducing piston width and weight as a result of axially narrow piston rings, but with a reduction in the overall width of oil rings being preferred to a width reduction in the compression rings to less than 1.5 mm.

REFERENCES

(1) JAKOBS
Zur Reibung der Kolbenringe bei PKW-Ottomotoren - Rechnerische Abschätzung der Reibungs- und Kraftstoffverbrauchsreduzierung - Ergebnisse motorischer Untersuchungen
GOETZE Fachschrift K 34, 1983

(2) FURUHAMA
Piston Ring Motion and Its Influence on Engine Tribology
SAE 790 860

(3) THIELE
Motorreibung
FVV-Heft Nr. 297, 1981

(4) MORSBACH
Einfluß der axialen Höhe von Kolbenringen auf deren Funktionsverhalten
MTZ 43 (1982), 7/8

(5) SPINDT
Air fuel Ratios from Exhaust Analysis
SAE 650 507

(6) SCHWEITZER, ZEILINGER
Das optimale Luft-Brennstoff-Verhältnis für Ottomotoren
MTZ 32 (1971) S. 16-21

Table 1 Table for all mathematically calculated measures for reducing friction
Gasoline engine, bore/stroke 85/86 mm

	Measure	Partial load, n = 2100 rev/min			Full load, n = 5000 rev/min		
		$\Delta P_R/P_R$ Single piston ring	$\Delta P_R/P_R$ Total engine	$\Delta b_e/b_e$ for η_{mo} = 0.45 p_e = const.	$\Delta P_R/P_R$ Single piston ring	$\Delta P_R/P_R$ Total engine	$\Delta b_e/b_e$ for η_{mo} = 0.80 p_e = const.
Compression ring	h = 2.0 → 1.5 mm	11.3 %	0.46 %	0.25 %	9.0 %	0.36 %	0.07 %
	h = 1.5 → 1.2 mm	6.9 %	0.28 %	0.15 %	6.0 %	0.24 %	0.05 %
	F_t = 10.1 → 8 N, h = 1.5 mm	6.2 %	0.25 %	0.14 %	6.0 %	0.24 %	0.05 %
	$\Delta F_t/F_t$ = 50 %, h = 1.5 mm	16.2 %	0.64 %	0.35 %	12.7 %	0.50 %	0.10 %
Oil ring	h_5 = 0.40 → 0.25 mm	3.7 %	0.44 %	0.24 %	2.8 %	0.34 %	0.07 %
	h_5 = 0.40 → 0.25 mm, p_o = 1.34 → 0.6 N/mm²	26.8 %	3.2 %	1.8 %	25.8 %	3.1 %	0.62 %

Table 2 Results of the tests with rings having a different tangential force F_t

	BMEP bar	n rev/min	$F_{t\,max}$	$F_{t\,min}$
Torque at motoring	4.32	3790	38.6 Nm	37.9 Nm
	1.14	2110	22.7 Nm	21.8 Nm
Lubricating oil consumption	1.84	5400	32 g/h	69 g/h
	5.01	5400	41 g/h	77 g/h
	full	5400	12 g/h	77 g/h
	test cycle		16 g/h	43 g/h
Blowby volume	full	5400	1.3 m³/h	1.4 m³/h
	0	5400	1.0 m³/h	1.6 m³/h
	0	5500	1.0 m³/h	1.9 m³/h
	0	5600	1.0 m³/h	2.1 m³/h

Table 3 Results of the tests with rings having a different axial width

	BMEP bar	n rev/min	Production ring pack	Axial narrow ring pack
Torque at motoring	full	3000	28.7 Nm	28.7 Nm
Lubricating oil consumption	1.84	5400	63 g/h	77 g/h
	5.01	5400	55 g/h	68 g/h
	full	5400	20 g/h	36 g/h
Blowby volume	full	5400	1.4 m^3/h	1.4 m^3/h
	0	5400	1.2 m^3/h	0.7 m^3/h
	0	5500	1.3 m^3/h	0.9 m^3/h
	0	5600	1.3 m^3/h	0.9 m^3/h

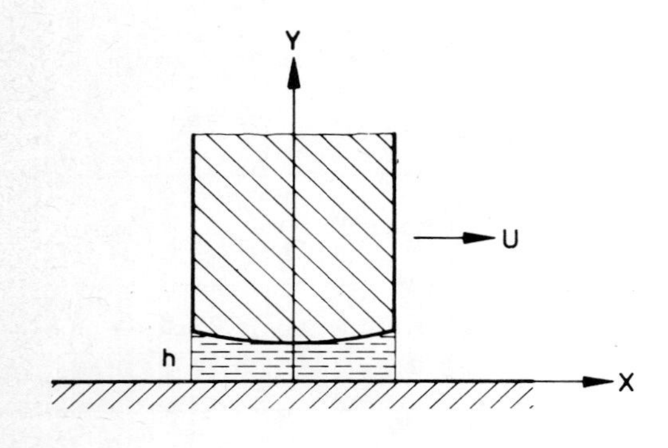

U = Piston speed
h = Width of lubrication gap
p = Pressure in lubricant
η = Viscosity
τ = Shearing stress

Reynolds equation

$$\frac{\partial}{\partial x}\left(h^3 \frac{\partial p}{\partial x}\right) = 6\eta\, U \frac{\partial h}{\partial x} + 12\eta \frac{\partial h}{\partial t}$$

Newton's shearing stress hypothesis

$$\tau = \eta \frac{\partial u}{\partial y}$$

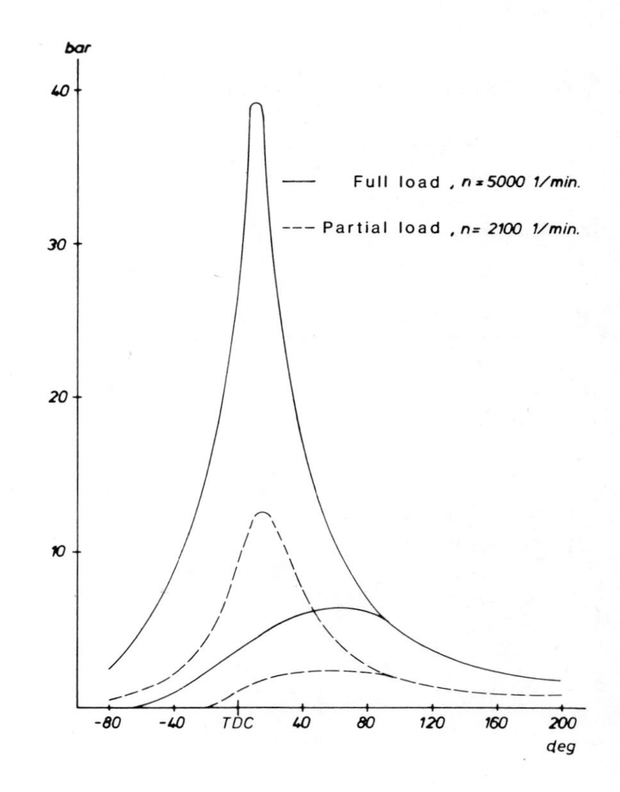

Fig 1 Reynolds differential equation
Newton's shearing stress hypothesis
Pressure curves above and below the top compression ring

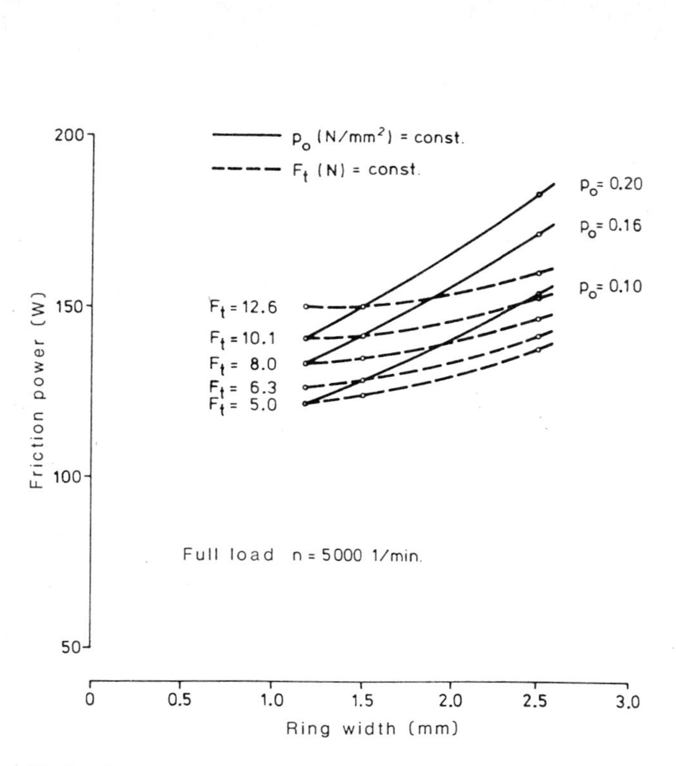

Fig 2 Friction power as a function of ring width.
Parameter: specific surface pressure, tangential force

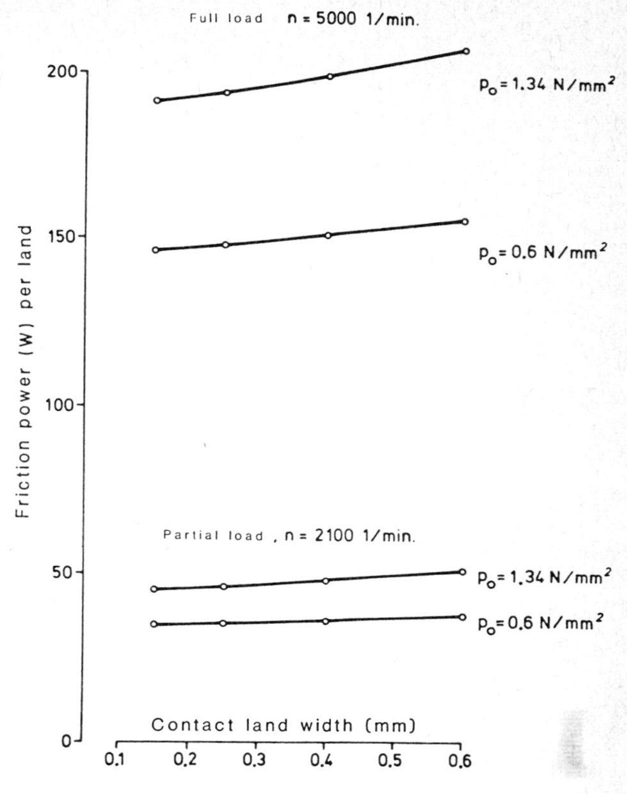

Fig 3 Friction power as a function of the contract land width of oil control rings.
Parameter: specific surface pressure

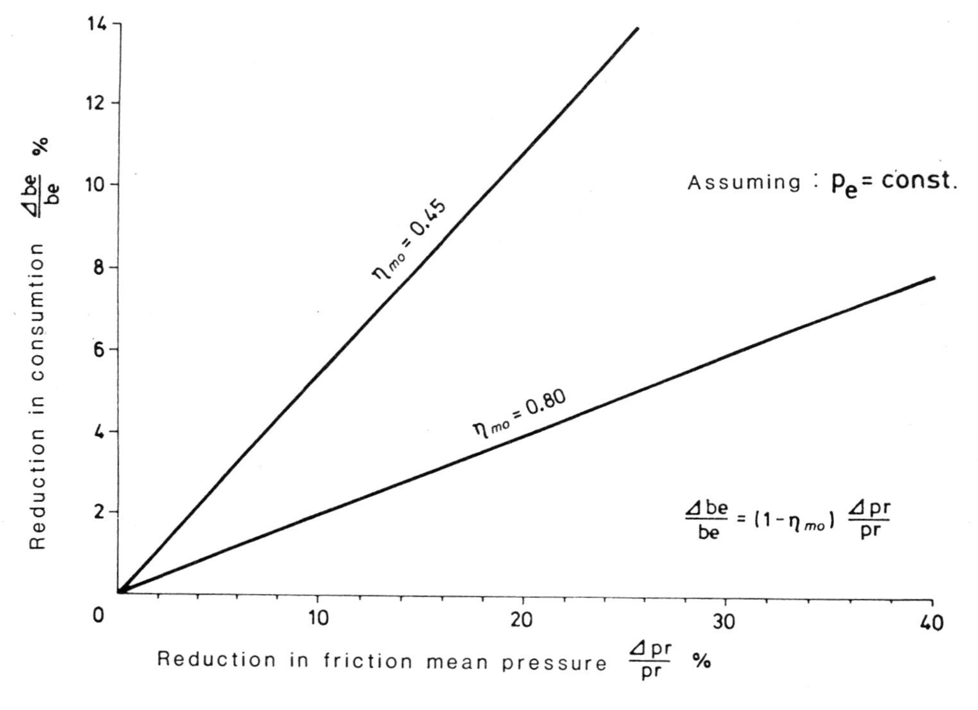

Fig 4 Reduction in consumption as a function of the reduction in friction mean effective pressure for p_e = constant

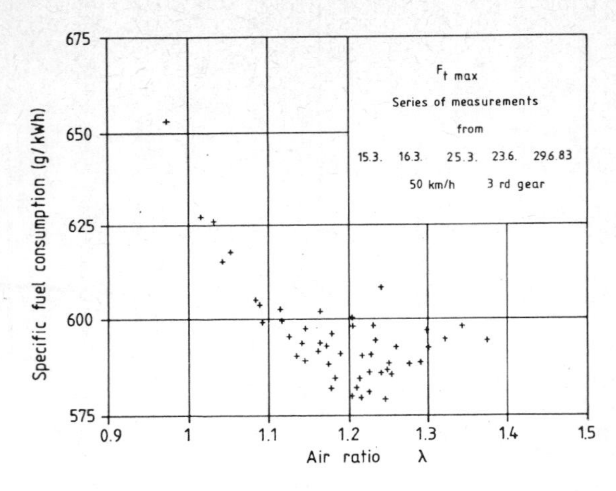

Fig 5 Scattering of BSFC — values measured within a period of four months

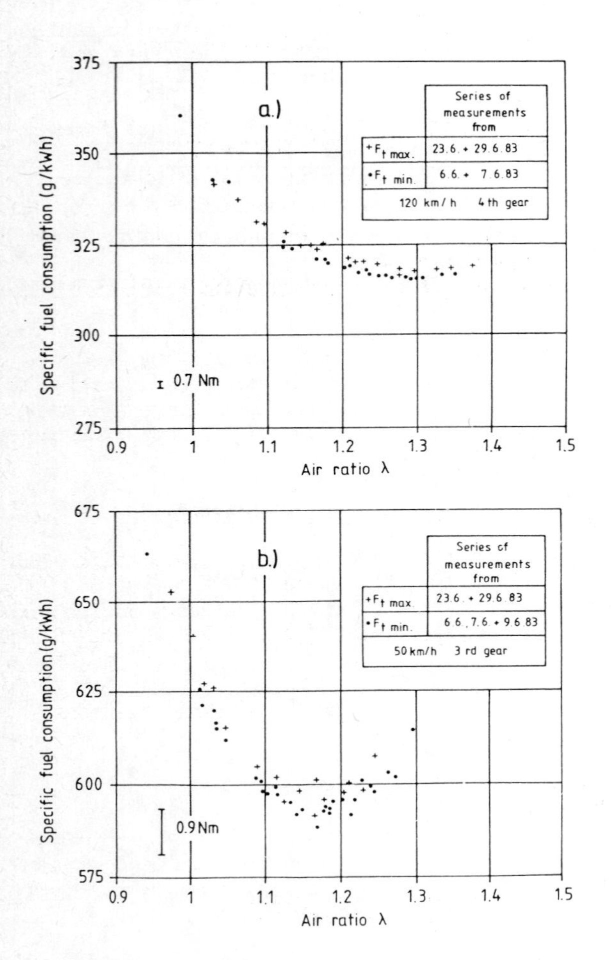

Fig 6 BSFC as a function of air ratio with rings having a different tangential force F_t

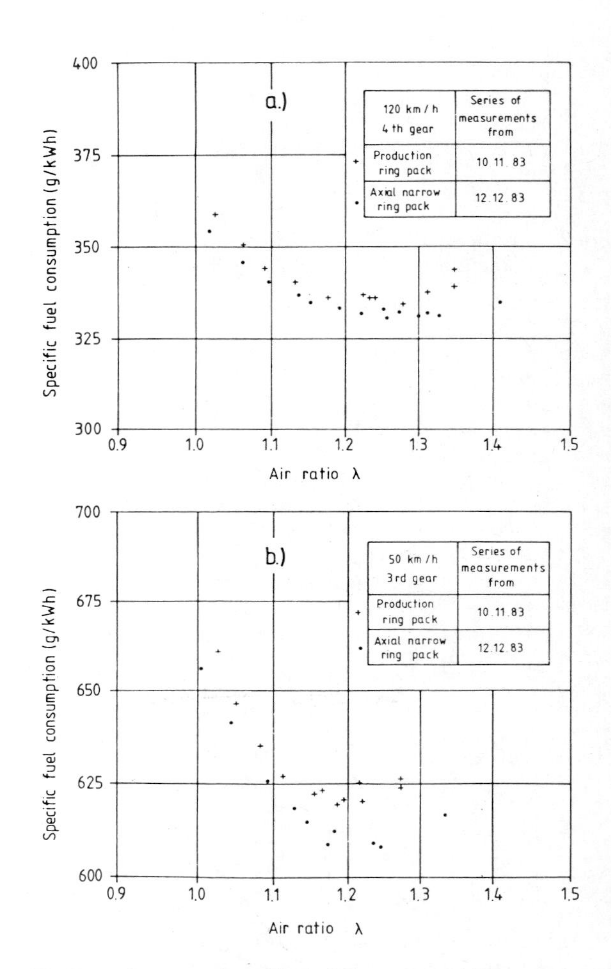

Fig 7 BSFC as a function of air ratio with rings having different axial width

C63/85

Development of a two-ring piston with low friction and small compression height without increase in blow-by, overheat and oil consumption

S F FURUHAMA, M H HIRUMA and H T TAKAMATSU
Musashi Institute of Technology, Tokyo, Japan
K SHIN
Nippon Piston Ring Company Limited, Yono-city, Saitama-Pref., Japan

Synopsis Improved two-ring piston has a superior or same performance compared with the conventional three-ring piston , on blow-by, piston temperature, piston frictional loss, oil consumptions and lower piston compression height.

1 INTRODUCTION

The two-ring piston was developed by Musashi I.T. Japan and reported at SAE International Congress 1984(1). Since then, further development and experimental work have been continuously carried out in cooperation with Nippon Piston Ring Corp. In this paper, a state of the art on the two-ring piston will be presented based on these experimental results.
In general, the two-ring piston has the following advantages;
(1) Decrease in piston frictional loss
(2) Decrease in piston compression height and engine total height
(3) Light weight and low cost piston
(4) As a result of the above advantages, an improvement of fuel economy and engine cost can be achieved.
On the other hand, there were some problems to be resolved such as increase in blow-by, piston crown temperature and oil consumption. The experiments have been carried out to overcome these problems. Then a improved two-ring piston has been developed. The new two-ring piston has the above mentioned advantages but does not suffer from the disadvantages.

2 EXPERIMENTAL METHOD

2.1 Test engine

The test engine is a four cylinder gasoline engine with a turbocharger for a passenger car and has the following specifications; 85 mm bore, 78 mm stroke, 1770 cm^3 volume of piston displacement, 8.0 : 1 of compression ratio. And its piston ring arrangement is as follows: the top ring of B = 2 mm and T = 3.5 mm is rectangular without chromium plate and its contact pressure Pe is 0.12 MPa. The second ring is the same as the top ring and the oil ring is a three-piece type.

2.2 Temperature measurement

Temperatures at 24 points on and inside the piston and 10 points on the cylinder bore have been measured simultaneously by the iron-constantan 0.15 mm thermocouple wires. The wires are led out along the link mechanism.

2.3 Measurement of piston frictional force

Instantaneous piston frictional force under firing engine operation was measured by the floating liner method(2).

2.4 Measurement of piston-ring axial movement inside the groove

The upper surface of each ring groove has an annular electrode in the entire circumferential direction. It forms the electric capacitance on the ring upper surface so that the ring movement can be detected(3).

2.5 Measurement of oil consumption

A continuous measurement of oil consumption has been carried out by the hydrogen fueled 0.3 ℓ mono-cylinder test engine. In this engine, carbon atom in the exhaust gases does not originate in fuel but only in lubricating oil(4). The conventional oil sump level method is used to measure the oil consumption in four-cylinder production engines.

2.6 Experimental procedure

All measurements were begun after the running-in operation was finished and the engine operating condition reached the steady state on the testing bench.

3 BLOW-BY CONTROL

One pressure ring piston naturally shows a larger blow-by than the conventional two pressure ring piston if the same rings are used. However, this disadvantage of two-ring piston can be overcome by means of a simple countermeasure, that is to use a non-chamfer ring.
In Fig.1-(a), type (1) is non-chamfer groove and ring, where a blow-by passage area at the gap A_1 is cg (about 0.06 mm^2) only. But type (2) is an originally used piston with chamfered groove and rings. The chamfer increases the blow-by passage by $2a_1 + 2a_2$ (about 0.14 mm^2) as shown (c).

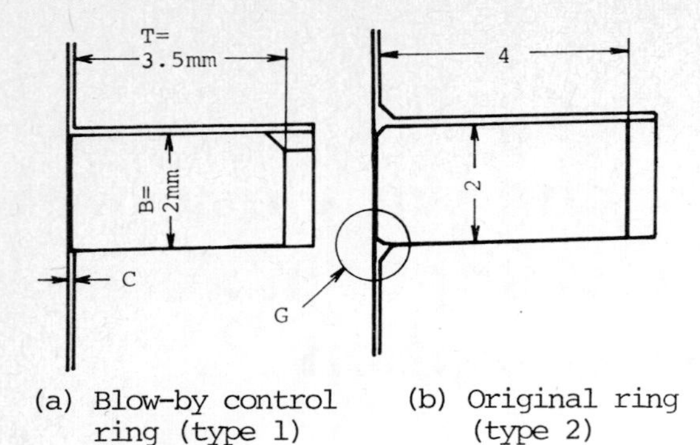

(a) Blow-by control ring (type 1)

(b) Original ring (type 2)

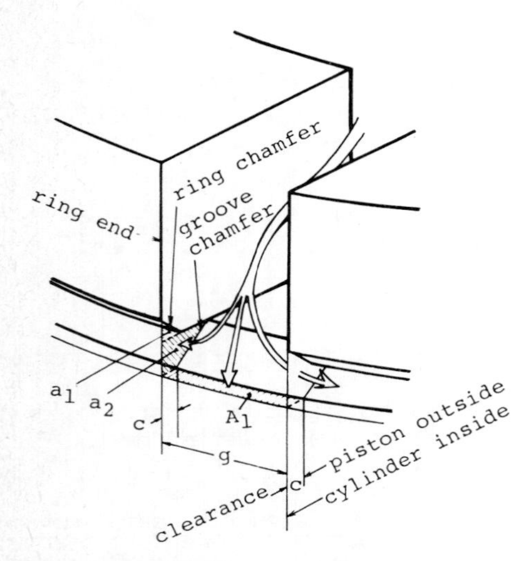

(c) Three dimensional illustration of blow-by gas passage at a ring gap

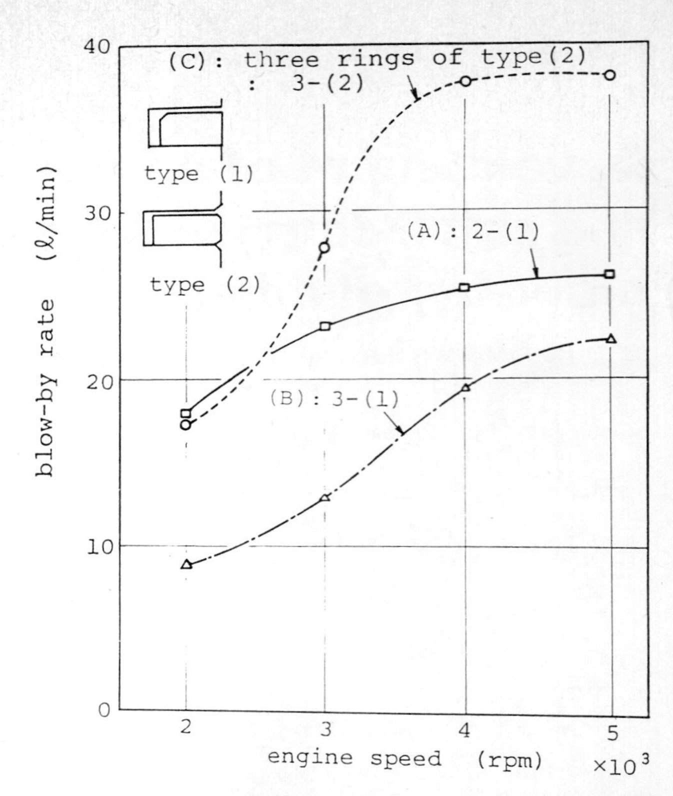

(d) Measuring result of blow-by characteristics under full load

Fig 1 Effect of chamfers on the ring and groove corner on the blow-by characteristics

As an empirical fact, it is known that the effective blow-by passage area at the periphery and under surfaces of the ring is approximately equal to the cg. Therefore a total area A_T can be estimated as follows:

Type (1) without chamfer

$A_T = 2A_1 = 2cg$ ($0.12\ mm^2$)

Type (2) with chamfers

$A_T = 2A_1 + 2a_1 + 2a_2$ ($0.26\ mm^2$)

Fig.1-(d) shows the measurement result of the blow-by rate. This result indicates the following important characteristics.

(1) Comparing the blow-by rate (A) of two-ring type (1) piston with (B) of three-ring (1) piston, the former rate is larger than the later one as one expects.

(2) Comparing the blow-by rate (B) with (C) of the three-ring type (2), obviously the former blow-by is about half of the later one because the type (1) has a smaller passage area A_T.

(3) Comparing (A) of the blow-by control two-ring piston with (C) of the original piston. Under the low speed, both pistons indicate about the same amount of blow-by. However, at high speed, (C) sharply rises beyond (A) because the second ring of the three-ring piston lifts up from the lower surface of the groove. The inertia force of the ring becomes larger than the pressure force, then the second ring lifts up and loses a sealing effect of gas leakage.

As a result, using a non-chamfer pressure ring, the blow-by control of two-ring piston can be improved and becomes better than that of the practically used chamfered three-ring piston.

4 CONTROL OF MAXIMUM PISTON TEMPERATURE

4.1 Test piston

Usually, about 70% of the heat inflow to the piston crown flows out into the cylinder through rings(5)(6). For the two-ring piston, the cooling may be insufficient and the piston can overheat. Actually, when the second ring was taken off from the original engine, knocking occurred and sometimes piston was seized. This problem has been worked out by changing the position of the rings to the upper positions. In each case, the same pressure ring type (1) and the same three-piece oil ring have been used. And each ring position was following.

(a) The three-ring type piston, denoted as a 3-ring type, had original groove position, but the grooves and the pressure rings had no chamfers.

(b) The two-ring type piston, the position and the type of oil ring were the same as the 3-ring type, and a pressure ring was located about middle of two pressure rings of the 3-ring type, then the top land length became 10 mm long. This type piston was denoted as 2-LT type (long top-land type)

(c) The two-ring type piston, which was the same type rings as the 2-LT type, but the ring

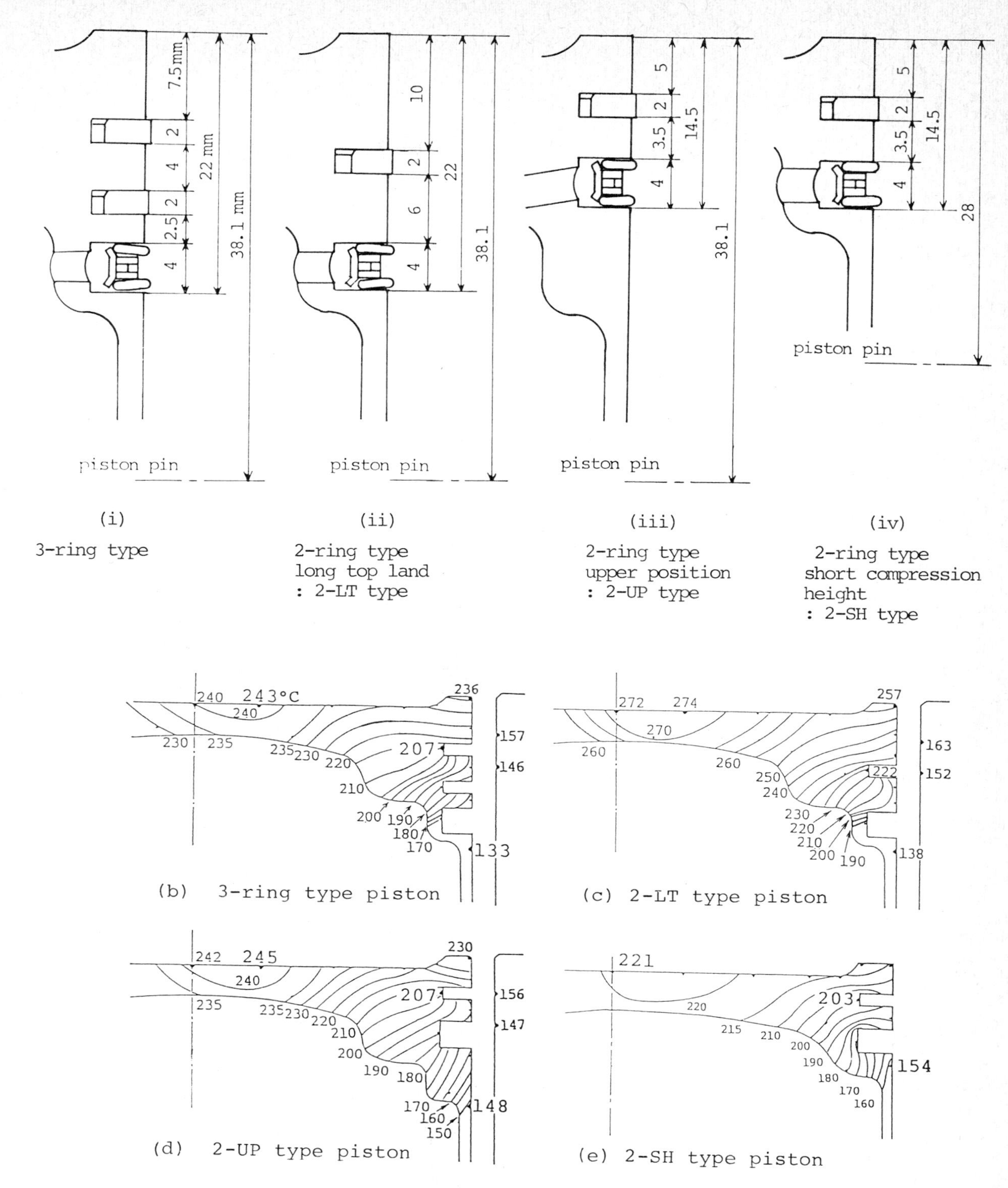

Fig 2 Measured temperature distribution of piston at 4000 r/min full load

positions were as high as possible. This type piston was denoted as 2-UP type (upper most position type).

4.2 Temperature distribution

The test engine as a turbocharged gasoline engine tends to cause the knocking at the wide open throttle operation using the regular gasoline (RON = 91). And the knocking limit of engine load restricts the maximum power of the engine. Therefore piston temperatures were measured just before the knocking limit load under each speed. Fig.2 shows temperature measurement results.

Firstly, comparing 2-LT type with 3-ring type, the temperature on the upper surface of the former type piston crown is clearly higher than that of the latter type by about 30°C under full load at 4000 rpm, then a spark timing had to be delayed for knocking free operation. As a result, as shown in Fig.3, the maximum power of the 2-LT type is lower than the 3-ring type because of a later spark timing than the optimum for the maximum power.
On the other hand, the temperature of the 2-UP type falls to about the same level as the 3-ring type, and its knocking tendency also falls near to that of the 3-ring type. And maximum power

of the 2-UP type is higher than that of the 3-ring type by 1.1KW which corresponds to 3.5% of the original output power at 2000 rpm. This gain is probably because of a reduction of the piston frictional loss.
An interesting fact is indicated on the temperature distributions. The upper surface temperature of 2-UP type piston crown becomes lower by heightening the ring positions, but the temperature at the upper part of piston skirt is raised. That is an effect of the high ring position, the axial temperature distribution on the exterior piston surface near the ring belt is leveled off.

4.3 Effect of shorter piston compression height or higher piston pin position

The 2-UP type has the advantages of shorter compression height i.e. lighter piston weight and shorter total engine height. This type piston was denoted as 2-SH (short compression height type) type. It is found that shortening compression height by 10 mm also makes the crown surface temperature lower as shown in Fig.2-(e). And it shows a higher performance as shown in Fig. 3 and Fig.5 because of a lighter piston.

5 REDUCTION OF FRICTIONAL LOSS

Instantaneous piston frictional forces were measured by means of the floating liner device. Its engine was mono-cylinder test engine with 85 mm bore, which was same as the above mentioned four cylinder test engine. The same piston as that of the four cylinder engine was assembled into the frictional force measuring engine.
Fig. 4 shows some examples of the measurement result. 2-UP type has a smaller friction force than the 3-ring type. A reduction of the friction

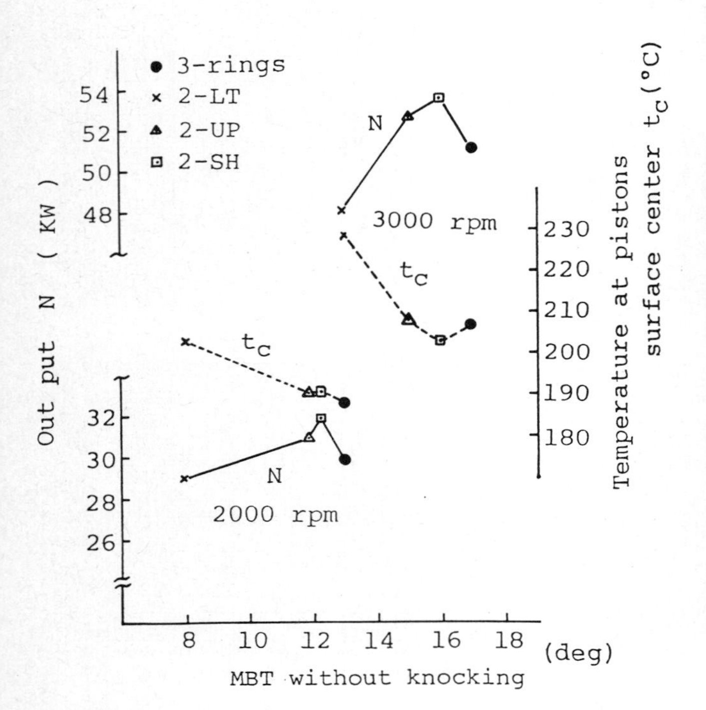

Fig 3 Maximum power without knocking of two-ring piston

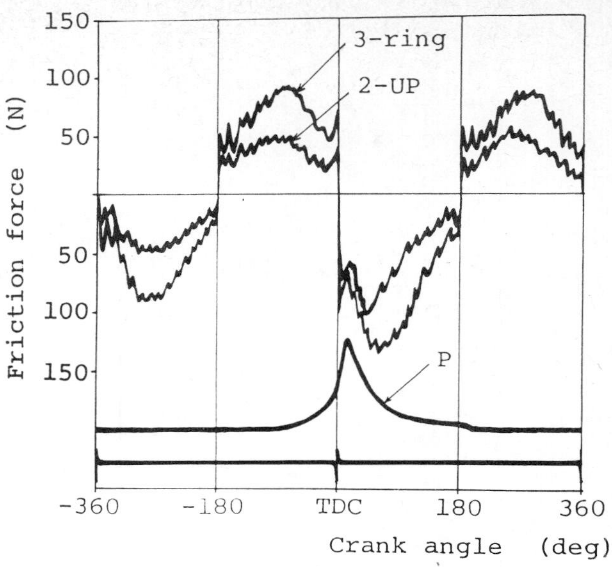

(a) Half load at 2000 rpm as a typical condition

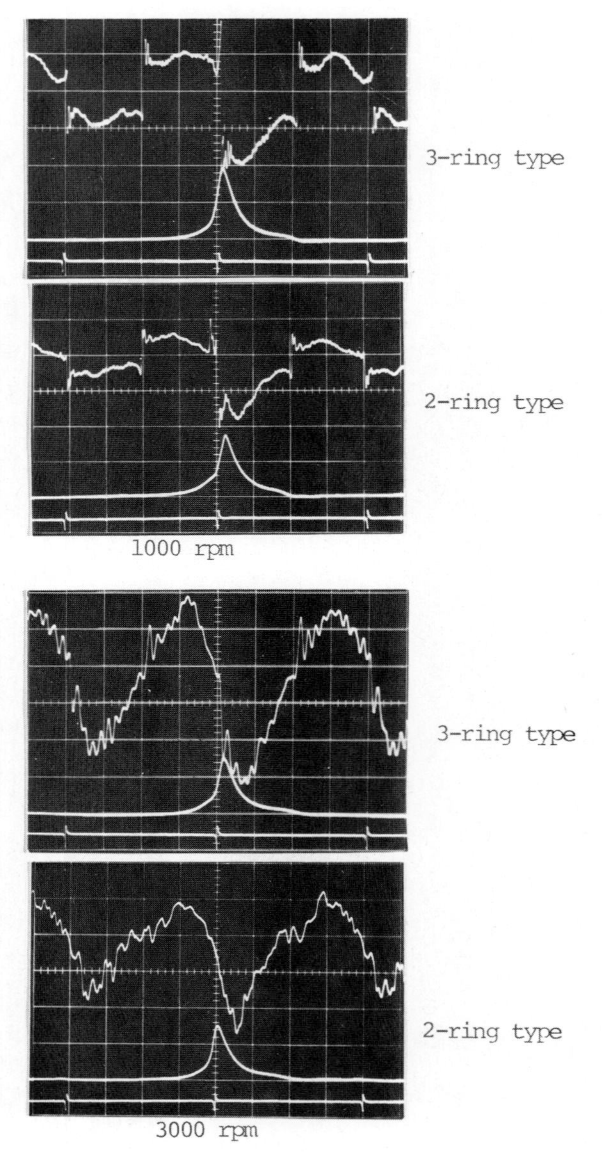

Fig 4 Comparison of measured piston frictional force of 2-UP type with three-ring type under 90°C at centre of top ring stroke on bore surface

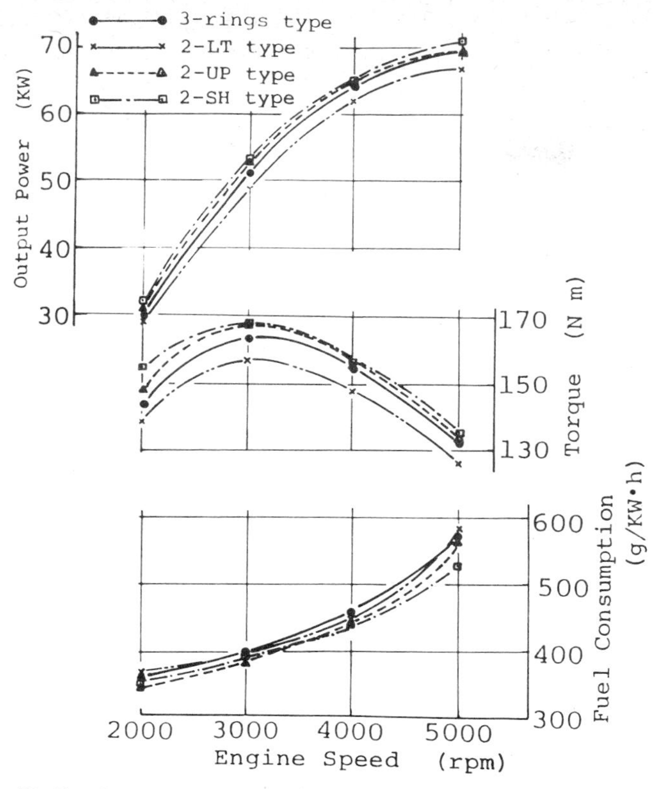

Fig 5 Comparison of engine performance using each type piston

loss amounts to about 40% of the 3-ring original piston. This effect seems too large as a result of removing one ring from three rings. It can be however understood, in the following way. 2-UP type reduces the oil starvation in the piston lubricating surfaces, and also the oil has a high temperature or low viscosity on the piston skirt. Then the over all engine performance is improved as shown in Fig.3 and 5.

6 PISTON RING AXIAL MOVEMENT

Since an axial movement of the piston ring in groove gives an important effect upon the blow-by and the oil consumption, it seems that the two-ring system has a different ring movement from the three-ring system because the lower side pressure on the top ring in two-ring system is approximate same as pressure in the crankcase. Main effects of the ring movement are shown in the photographs of Fig.6. And these can be indicated as following.

(1) If the lower side pressure of the top ring p_2 is higher than crankcase pressure, and p_2 becomes equal to the combustion chamber pressure p_1, the top ring lifts up as shown ① in Fig.6-(a-1). Then the gas with pressure p_2 is blown up into the combustion chamber accompanied by a lubricant oil. As a result, the blow-by reduces , but the oil consumption will be increased in case of 3-ring type. On the other hand, the top ring of 2-UP type never causes that phenomenon as shown in Fig.6-(a-2).

(2) At low load, the second ring of the three-ring system tends to lift up as shown ② in Fig. 6-(b-1). Then the gas with pressure p_2 is blown

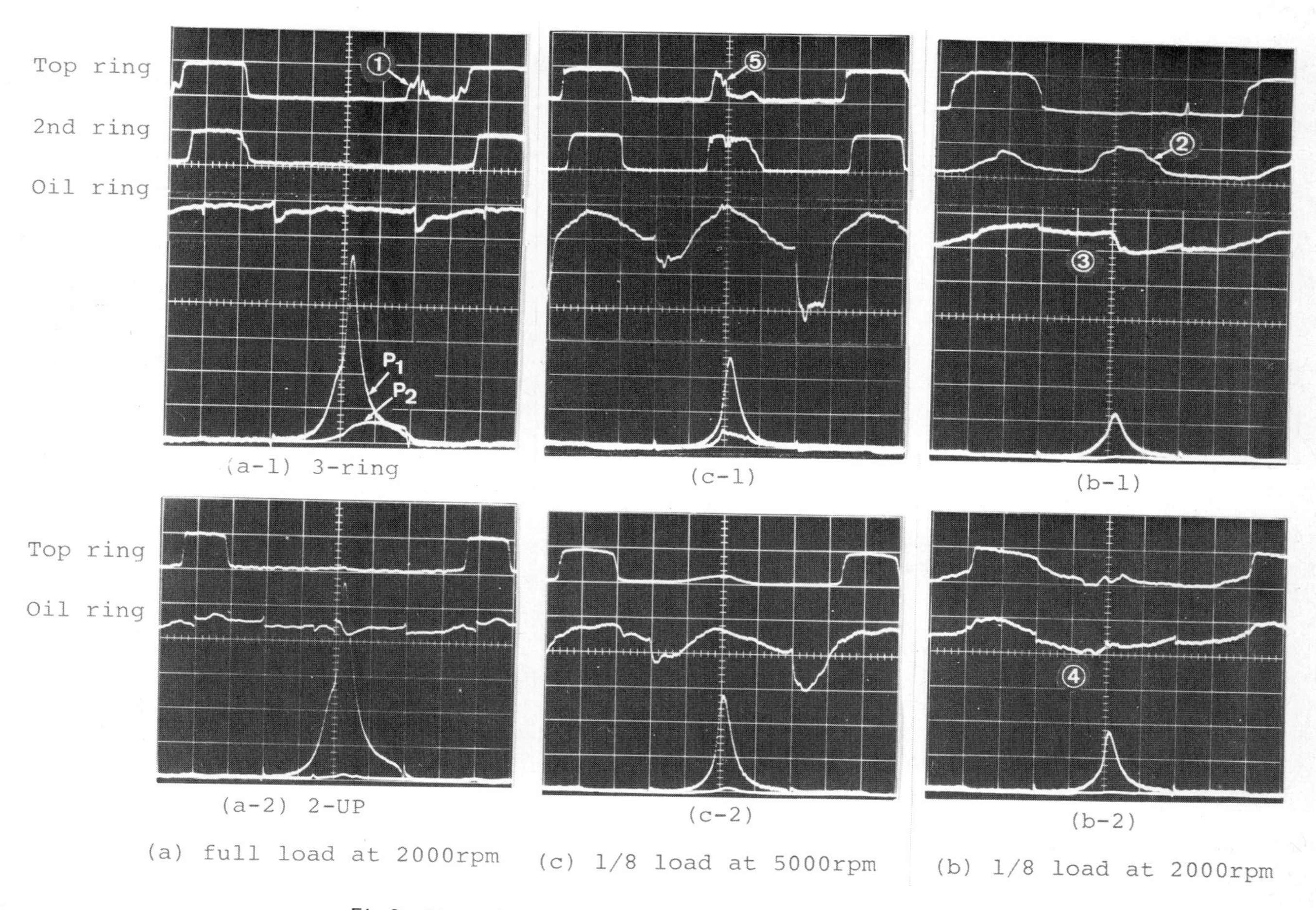

Fig 6 Measuring the result of piston ring axial movement

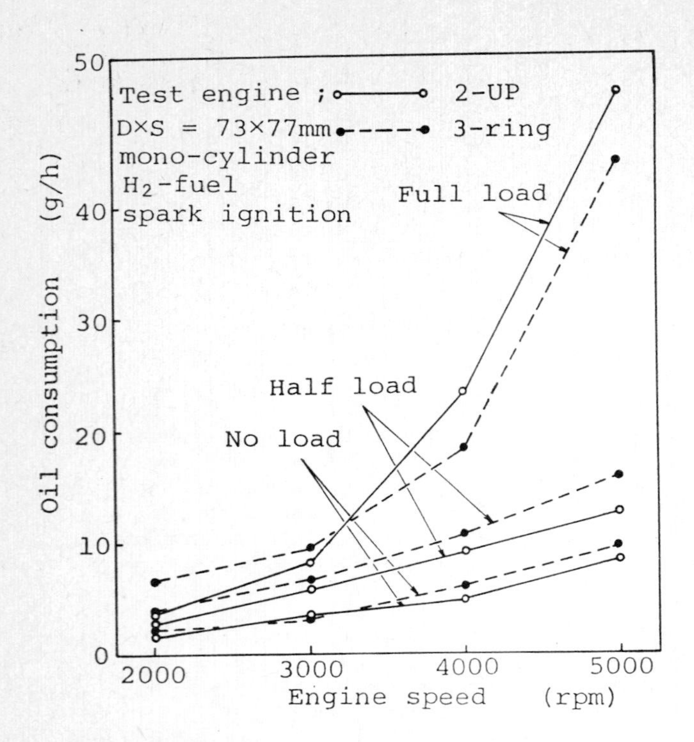

Fig 7 Comparison of 2-UP type oil consumption with three-ring type in hydrogen fueled engine

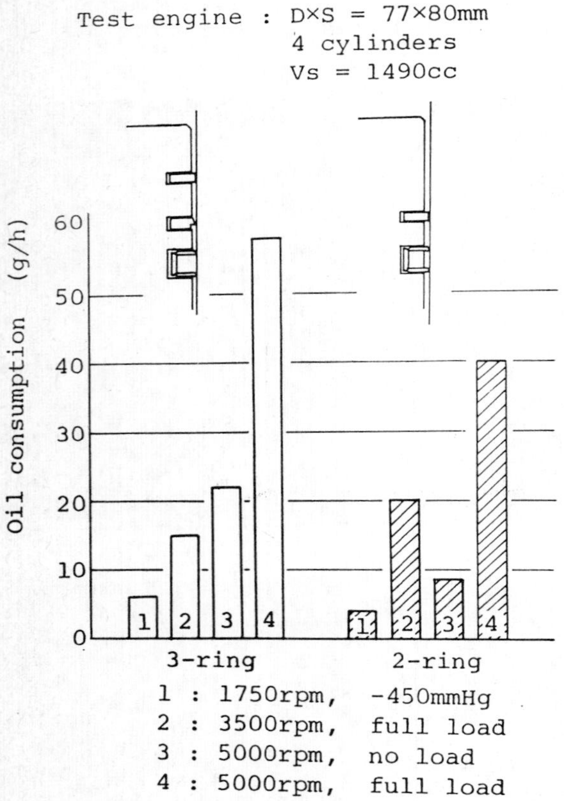

Test engine : D×S = 77×80mm
4 cylinders
Vs = 1490cc

Oil consumption (g/h)

3-ring 2-ring

1 : 1750rpm, -450mmHg
2 : 3500rpm, full load
3 : 5000rpm, no load
4 : 5000rpm, full load

Fig 8 Comparison of oil consumption between three-ring and two-ring in a production engine

down into the crankcase accompanied by the oil. As a result, the blow-by gas increases but the oil consumption will be reduced.
At that operating condition, pressure in the cylinder is low, and in some parts of a cycle, it becomes lower than crankcase pressure. Then the ring movement is not stable, especially in case of the two-ring system as shown in Fig.6-(b-2). Therefore the oil consumption will be increased.

(3) Ideal movement of the oil ring is such that the ring movement is governed by the ring friction force only. From this point of view, the oil ring movement of 2-UP shown as ④ in Fig.6-(b-2) is better than that of 3-ring type shown as ③ in Fig.6-(b-1).

(4) At low load and high speed, the top ring of the 3-ring type lifts up at the same time as the second ring does as shown ⑤ in Fig.6-(c-1). This is a harmful phenomenon because neither pressure rings play the gas sealing action. Sometimes it leads to the scuffing on the ring and the cylinder surfaces. On the other hand, in two ring system, the top ring does not indicate such harmful phenomenon because p_2 is very low.

7 OIL CONSUMPTION CONTROL

Using the two-ring piston in practice, one of the most important problems seems a control ability of the oil consumption. The above mentioned experimental result indicates, on the whole, that the 2-UP type has rather superior movement than the 3-ring type for the oil consumption control. On the other hand, it is expected usually that an oil consumption of the two-ring piston engine may be larger than the three-ring piston engine because the number of rings is small.
Contrary to one's expectation, the measurement result indicates that the control of the oil consumption by the 2-UP type is not inferior to the 3-ring type. Fig.7 shows a comparison between the 2-UP type and the 3-ring type under each operating condition in the test engine using hydrogen fuel. At one time in this series of experiment, it was experienced that the 2-UP type indicated a larger oil consumption than the 3-ring type. This was a special phenomenon because the cylinder bore was unusually deformed. Next, Fig.8 shows an another result to indicate the same matter, which was taken by the practical production four cylinder engine.
These results indicate that an oil consumption control ability of the 2-UP type is about the same level as the 3-ring type, because the ring movement of the former is more favorable than the latter.

8 CONCLUSION

From these experimental results, the following main conclusions are obtained;

(1) Blow-by of the two ring piston can be controlled to a sufficiently low level by using a pressure ring with good sealing ability, for example, the non-chamfers on the ring and the groove corners.

(2) When two rings are set at the uppermost position (2-UP type), the temperature on the upper surface of the piston crown decreases to the same level of the three-ring piston, and the knocking limit is also

controlled. However, the temperature on the upper part of the piston skirt rises.

(3) 2-UP type piston can shorten the piston compression height (heighten the piston pin position) by 10 mm. Its maximum temperature decreases further and engine performance is improved by reduction of the piston weight.

(4) Frictional force of the 2-UP type piston is smaller than that of the 3-ring type piston by 40% of the latter. And overall engine performance is improved by about 3.5% at 2000 rpm.

(5) Axial movement of the ring in the 2-UP type is different from the ring in the 3-ring type. And the former ring has rather superior movement than the latter for the oil consumption control.

(6) On the oil consumption under the normal engine condition, the 2-UP type is about same level of the 3-ring type because the former rings have the favorable movement for the oil consumption.

9 REFERENCES

(1) FURUHAMA, S. KOJIMA, M. ENOMOTO, Y. YAMAGUCHI, Y. Some study on two-ring piston in automobile turbocharged gasoline engine. SAE paper 840183, 1984

(2) FURUHAMA, S. SASAKI, S. New device for the measurement of piston frictional force in small engines. SAE paper 831284, 1983

(3) FURUHAMA, S. and HIRUMA, M. Axial movement of piston ring in the groove. ASLE Transactions, Vol. 15, 4 , p278 - 287, 1972

(4) FURUHAMA, S. and HIRUMA, M. Some characteristics of oil consumption measured by hydrogen fueled engine. Lubrication Engineering, Vol.34, 12, p665 - 675, 1978

(5) TADA, T. and FURUHAMA, S On the heat flow from the piston in a farm type gasoline engine. Bulletin of JSME Vol 7, 28, p784 - 791, 1964

(6) FURUHAMA, S. and SUZUKI, H. Temperature distribution of piston rings and piston in high speed diesel engine. Bulletin of the JSME, Vol. 22, 174, p1788 - 1795, 1979

C70/85

The AEconoguide low friction piston feature – analysis and further running experience

D A PARKER, BSc, PhD, ARCS, CEng, FIMechE, FInstP,
C M McC ETTLES, BSc, PhD, DSc, ACGI, DIC, CEng, MIMechE and
J W RICHMOND, BTech
AE Developments Limited, Cawston, Rugby, Warwickshire

SYNOPSIS The operating principles of the design are described, namely the use of typically three discrete pads to restrict piston to bore contact area to an extent consistent with a reduction of viscous loss, combined with the establishment of good skirt guidance. The design is suitable for hydrodynamically lubricated bearing surfaces which operate on the high film thickness side of the friction minimum in the Stribeck curve, care being necessary to ensure that the minimum is not approached even under the most adverse operating conditions. Improved guidance is obtained by the use of discrete contact areas at the upper and lower extremities of the skirt.

The basic theory for a single pad is presented, allowing the calculations of viscous loss as a function of angle of inclination between the pad and bore. Measurements of piston transverse movement within the bore are presented and comparisons provided between traditional and AEconoguide designs. The theoretical treatment is then extended in the light of the changes of piston inclination and position due to transverse movement within the bore, so as to provide overall predictions of viscous loss. Finally, a summary is given of results obtained in running engines, both as a statement of the contribution of the design to the overall reduction of engine friction, and for comparison with the theoretical results.

(1) INTRODUCTION AND OPERATING PRINCIPLES

The AEconoguide principle was conceived as a means of reducing piston friction by substantially reducing the skirt bearing area, whilst maintaining or improving effective lubrication. There were two intrinsic prerequisites, firstly that special piston castings would not be needed and secondly, that machining would be performed on existing numerically controlled diamond turning lathes developed by the AE Group. The ultimate target was, and is, to reduce skirt friction by a significant proportion and either harness the energy thus released as extra power or use the improved efficiency to save fuel.

The mechanical losses in a conventional piston engine usually represent less than 10% of the overall losses but a much greater proportion of the power output, falling from 100% at idle to around 20% at high power. Friction losses incurred by the piston skirt alone constitute 10-20% of the mechanical losses and therefore 10-20% of the output at idle falling to 2-4% at high power (Ref. 1). Consequently, if piston friction were halved, the energy regained would raise power output and thus overall efficiency by 1-10% depending upon operating conditions; the original power could then be restored by reducing fuel in approximately the same proportion. Where road loads are low, e.g. when speed restrictions apply, the savings will tend to be biassed towards the upper end of this range.

The lubrication of piston skirts is essentially hydrodynamic; not only has this been demonstrated by the measurement of oil film pressures (Ref. 2), but if it were not so the wear rate and coefficient of friction associated with piston skirt to bore contacts would be much higher than the observed values.

Such boundary lubricated contact as does occur between these surfaces arises during running-in when the surfaces are progressively smoothed. As running-in continues the proportion of the cycle associated with boundary lubrication decreases steadily, with the hydrodynamic regime spreading from mid-stroke towards the dead-centre positions. Thus, for a well run-in piston, boundary lubrication is only likely to occur on start-up, around the dead centres,and when the contact conditions are modified by a sudden change in speed or load.

Compared with many journal bearings and with piston rings, the piston skirt is relatively lightly loaded; decreasing the effective contact area between piston and bore therefore offers a potential means of reducing piston friction. There is an important proviso, however, which is that in doing so the film thickness must not be reduced to a value at which boundary contact can occur. Thus the principle of reducing friction by decreasing contact area is applicable only to hydrodynamic bearings operating significantly to the right of the minimum friction point in the Stribeck curve (Fig. 1). Such conditions have been shown to be fulfilled in the lubrication of piston skirts.

In the AEconoguide piston design, of which an example is shown in Fig. 2, the reduced skirt contact area is sub-divided into a number of discrete bearing 'pads', typically raised 35μm above the remainder of the piston skirt. This value was chosen as sufficient to confine the load-bearing oil films between the piston and bore to the pad areas. Thus the only shearing of oil to be permitted was that which contributed directly to hydrodynamic support of the load. The number and disposition of the pads over the skirt area offered the possibility of a very large number of combinations, but the most usual design of 2 pads on the upper skirt and 1 on the lower was chosen to achieve stable contact at points as far distant from each other as practical.

It follows from the film thickness considerations given above that if the dangerous minimum friction point of the Stribeck curve is not to be approached, the pad area must be fully utilised.It is therefore particularly important in the AEconoguide design to ensure that the pad faces are in uniform contact with the bore. As the piston moves up and down in the cylinder bore, it executes a very precise and repeatable set of transverse motions within the bore (Ref.3) under the influence of gas pressure, inertia, friction force and varying angles of connecting rod obliquity. This tranverse piston motion results in inclination of the piston axis relative to the bore, again following a precise cyclic pattern which itself depends upon engine temperature, speed and load.

(2) THE THEORY FOR A SINGLE PAD

2.1 The General Calculation Scheme

At this stage in the development of the AEconoguide design it had been shown by the best possible method, experimental measurement in operating engines, to be effective (Ref. 4). However, there were many design parameters, all infinitely variable, which needed to be optimised. These included obvious dimensions such as pad size,location, number, relief, ramp angle and ramp extent. Other factors include piston clearances,taper,ovality,surface finish, piston-to-cylinder oil film thickness, the consequences of partial oil starvation and the mechanism of heat transfer between piston and cylinder.

Since there is clearly a limit to the extent to which such factors can be investigated experimentally, it was decided to construct a three-dimensional model of the AEconoguide features, to consider the motion of the skirt bearing these features within the cylinder and hence to calculate the overall contribution to cylinder friction attributable to the piston skirt during actual operating cycles. As a first step a single pad was considered to be operating at steady load and piston-to-cylinder inclination, although the load and inclination were varied between calculations (Ref.5).

2.2 Pressure and Temperature Distributions in the Oil Film Volume - Methods of Calculation

The basic representation used for pad geometry is shown in Fig. 3. One of the first matters investigated was the distribution of temperature over the film volume. This was given some priority, because one explanation offered initially for the mechanism of friction reduction was that concentration of the load into 3 relatively small areas raised the local viscous energy dissipation and hence the temperature significantly. It was thought that the consequent reduction in viscosity might account for the observed reduction in friction.

To investigate the matter the temperature was assumed to vary in three dimensions. However, to determine the temperature distribution it was necessary to solve simultaneously the pressure distribution; the latter was accomplished as follows.

If the film thickness is arbitrarily defined in two dimensions and the film temperature arbitrarily defined in three dimensions, then the generated pressure may be found by developing the stress equations (1, 2).

$$\frac{\partial p}{\partial x} = \frac{\partial \tau_x}{\partial z} = \frac{\partial}{\partial z}\left(\eta\frac{\partial u}{\partial z}\right) \qquad (1)$$

$$\frac{\partial p}{\partial y} = \frac{\partial \tau_y}{\partial z} = \frac{\partial}{\partial z}\left(\eta\frac{\partial v}{\partial z}\right) \qquad (2)$$

Here x is the direction of sliding
y is transverse to sliding (circumferential in piston)
z is through thickness of the film
η is local viscosity = $f(x,y,z)$
u,v are local velocities in x,y directions.
τ_x and τ_y are shear stresses.

Integrating equations (1, 2) twice and satisfying the boundary condition at $z = 0,h$ gives the velocity profiles $u(z)$ and $v(z)$ across the film. The velocity profiles may be integrated to give the local flows, q_x and q_y in the co-ordinate directions. Continuity may then be applied in the form

$$\frac{\partial q_x}{\partial x} + \frac{\partial q_y}{\partial y} = 0 \qquad (3)$$

This yields an equivalent of Reynolds equation for pressure,but without the usual restriction that the viscosity must be uniform through the thickness. The viscosity and film thickness must be defined for the Reynolds equivalent equation to be formed. These definitions may be arbitrary. The integrals:

$\int_0^z 1/\eta.dz$ and $\int_0^z z/\eta.dz$ appear and must be obtainable. Effective turbulent viscosities η_x and η_y were used in the analysis since representative Reynolds numbers were unknown. It was found that in the general case the Reynolds number was of order 20, so that superlaminar effects were negligible. Reynolds equation or its equivalent was solved by finite difference substitution and successive over-relaxation.

The solution method for temperature used was that of cell discretisation, in which the film is divided into inter-connecting, six-sided cells of trapezoidal cross-section when viewed normal to the direction of sliding (Fig. 4A). The eight corners of each cell are called grid nodes. At the geometric centre of each cell (in 3-D space) there exists a cell node. Lubricant enters or leaves a cell through each of the six faces (unless a face happens to be a film boundary). Conductive heat transfer is considered only in the direction through the film; convective heat flux is considered by assigning a temperature to flow into or out of each cell. For the sake of stability, this temperature is taken as being that of the (upstream) cell from which the flow occurred. The temperature at entry must be specified. This was taken as being uniformly at the cylinder wall temperature. The exit temperature profile to the sides and trailing edge are dependent parameters. To obtain a solution for temperature the film thickness and pressure fields must be completely defined, since the local velocities U and V (Fig. 3) must be defined, and these are pressure and film thickness dependent. The overall solution procedure involves successive solutions of the pressure and temperature distributions until both converge.

2.3 Temperature Distribution in Oil Film Volume - Results

It was decided to begin with films of very simple shape, partly because simplified classical analysis is possible for the simpler shapes and also because it was thought that the resolution of the multiple regression method might coarsen as features such as ramps were added.

The case that follows is for a plain, inclined wedge of film extent: B = 15mm, L = 15mm (Fig. 3) with a centreline convergence of 3:1. A non-dimensional flare of δ = 1.01 is added. An 8 x 8 x 8 mesh is used (Fig. 4A). The independent parameters and dependent parameters (results) are shown in Table 1 and the corresponding film thickness field in Table 2. Fig. 4B gives a contour plot of the film temperatures. The first column of cells (K = 2) lie forward of the leading edge. They exist only to define the entry temperature profile. The entry temperature is set to be uniformly equal to the wall temperature at 100°C.

A strong feature of the results is that the film temperature is governed almost completely by the values assigned to the pad and wall, 140°C and 100°C respectively. This can be seen from the contours of Fig. 4B which are almost evenly spaced. Proceeding from the leading edge in the downstream direction it can be seen that the temperature rises rapidly to a value linearly related to its distance from the wall. This occurs due to conduction effects and because the film is very thin. The calculated friction power loss was 18.75W compared with the conducted heat flux across the film of 163W. It is thus evident that the temperature distribution is dominated by conduction of heat across the oil film from the pad to the cylinder and not by generation of heat within the oil film. It would thus appear that the observed reduction in piston skirt friction is due to limiting the contact area per se rather than to any local heating arising therefrom.

2.4 Friction and Oil Film Thickness - Simplified Analysis

Now that the thermal conditions had been established, the task of calculating the oil film thickness and friction was greatly simplified. It was decided from the outset to correlate the calculated results by linear multiple regression. However, in order to give guidance in the choice of independent parameters for multiple regression a simplified analysis was undertaken. Ref.6 gives the non-dimensional load capacity of plain, inclined pads as a function of convergence κ and L/B ratio. The film efficiency reaches a peak at around κ = 1.125 and it was found necessary to curve fit the results separately for $\kappa < 1.125$ and $\kappa > 1.125$. The results of curve fitting for coefficient of friction and minimum film thickness were:

$\kappa < 1.125$

$$\mu = 2.57 \left[\frac{\eta U}{W}\right]^{\frac{1}{2}} \frac{B^{.45} L^{.05}}{\kappa^{0.35}}$$

$$h_o = 0.248 \left[\frac{\eta U}{W}\right]^{\frac{1}{2}} B^{.55} L^{.95} \kappa^{0.125}$$

$\kappa > 1.125$

$$\mu = 2.60 \left[\frac{\eta U}{W}\right]^{\frac{1}{2}} \frac{B^{.45} L^{.05}}{\kappa^{0.35}}$$

$$h_o = 0.267 \left[\frac{\eta U}{W}\right]^{\frac{1}{2}} B^{.55} L^{.95} \kappa^{-0.20}$$

2.5 Friction and Oil Film Thickness - Multiple Regression

It appears that μ varies monotonically with κ since the expressions for both regimes are almost identical. The exponents of h_o are the same in both regimes except for κ. These initial results indicated attempted regressions of the following form might give reasonable results.

$$\mu \text{ or } h_o = C\, B^a L^b W^c U^d \eta^e \kappa^f$$

The values of the constant C and the exponents a.......f may be found by standard methods providing there are seven results. The method of linear multiple regression allows many more results to be used. All of the variables are varied simultaneously in a random manner from case to case. The method thus incorporates a high degree of redundancy; for example a set of say 20 results may eventually be assembled as a set of 7 linear simultaneous equations to allow solutions for the constant and exponents. Many library routines are available for this purpose, the better of which give an indication of the confidence limits for the calculated data.

To obtain appropriate middle range values for the multiple regression a standard 4 cylinder engine of 70mm bore and stroke was considered. At 3000 rev/min the maximum sliding velocity Umax is approximately 11 metres/second.

Accordingly the speeds considered were randomly generated in the range 2 - 18 metres/second. Similarly for an instantaneous mean effective pressure of 1.00 MPa, the side thrust is 785N for a connecting rod obliquity of 20°. If the piston projected area (say 70mm x 70mm) is cut away by 85%, then a typical mean film pressure in the pads is 1.07 MPa and a typical pad size is B = L = 15.6 mm. From these and similar considerations the minima and ranges of the variables for multiple regression were taken as in Table 3.

2.6 The Effects of Flare, Inlet and Outlet Ramps, and Squeeze

The initial work considered plain converging wedge films but later this was extended to include flared converging wedge films to which were subsequently added inlet and outlet ramps. The nature of each of these configurations is indicated in Fig. 3.

In Table 4 the regression values for μ, the coefficient of friction, are given in respect of each of these configurations and compared with the corresponding values for the simplified analysis. The corresponding results for minimum film thickness are given in Table 5. The errors quoted in Tables 4 and 5 are the 'standard' errors given by the regression program and are about one half the corresponding r.m.s. errors. The calculations illustrate one important advantage of ramps in addition to that of ensuring convergence under all conditions. This is that for any given thickness of the oil layer in front of the pad the film length will adjust itself beneficially with changing load. For example, an increase in load would tend to cause an increase in film length due to the reduction in minimum film thickness. This provides a stiffening effect which helps to avoid film breakdown.

A similar regression analysis was undertaken in respect of squeeze film lubrication experienced as the pads move radially towards the cylinder wall. Cases were considered both in which the whole of the pad moved radially outwards and also in which the pad was subject to simultaneous tilting, so that part moved radially outwards and part radially inwards.

However, even at the highest squeeze velocities considered in the analysis, the effect of squeeze under high film thickness conditions was to decrease the coefficient of friction by less than 3%. At low film thicknesses (i.e. around the dead centres) the effect rose to 11% but the corresponding effect on the integrated cyclic power loss would be much smaller. Accordingly, in the interests of simplicity, squeeze effects were omitted from the calculations described in Secion 4.

(3) THE IN-CYCLE VARIATION OF PISTON-TO-CYLINDER INCLINATION

The positive clearance of a piston in its cylinder allows lateral movement to occur. The exact nature of this movement is the result of a complex interaction between many variables including speed,engine geometry, cylinder pressure, piston geometry and friction. Turning moments applied to the piston about the gudgeon pin axis, for example due to gudgeon pin friction or gudgeon pin offset combined with gas pressure, result in a tilting of the piston within the clearance space of the cylinder. However, under given engine operating conditions the pattern of piston transverse movement and its inherent tilt is repeated very accurately from engine cycle to engine cycle. Piston transverse movement has been measured (Ref. 3) using inductive displacement transducers mounted in the piston skirt. With such transducers mounted at the upper and lower extremities of the centre of the skirt on the thrust and non-thrust faces, it is possible to deduce each radial clearance and hence the piston axis inclination as a function of time throughout the operating cycle. Fig. 5 gives examples of such traces taken recently in the same engine fitted firstly with a standard monometal piston and secondly with a monometal piston with the AEconoguide feature.It will be seen that although the pattern is modified slightly, the head of the piston still moves across the entire clearance space which is, of course, many times the oil thicknesses on the piston and liner. It is thus evident that the load on the skirt of an AEconoguide piston may be thought of as being borne by one or more hydrodynamically lubricated wedges according to the inclination of the piston relative to the bore.

During parts of the cycle, for example the power stroke, both top and bottom of the piston skirt are in contact with the thrust face of the cylinder and the piston inclination remains near zero. This is described as Mode 1 operation. In Modes 2 and 3, the piston is tilted with the upper or lower part of the skirt respectively in contact with the bore. To model AEconoguide operation we must therefore consider the lubrication of each of the pads as they come in contact with the bore but remember that between such periods of contact, the non-contacting parts of the skirt, including the pads thereon, would be surrounded by air or oil mist.

The regular cyclic nature of piston movement lends itself to calculation using a step-by-step process throughout the operating cycle. In Ref. 7 the correlation between measured and predicted piston movement is shown to be good, so that for many purposes it is no longer necessary to make experimental measurements. Such calculations are used in Section 4, in conjunction with the theory developed for a single pad at a given inclination (Section 2), to calculate the frictional losses and oil film thicknesses associated with AEconoguide pads operating on realistic piston-to-cylinder inclination cycles.

(4) CALCULATED PERFORMANCE OVER AN ENGINE OPERATING CYCLE

The data derived from the piston movement program were the loads W1 and W2 on the upper and lower pads respectively, and the angle of inclination of the piston axis, each as a function of crank angle. When the piston was operating in Modes 2 or 3, i.e. with just the upper pads or the lower pad in contact, calculation of the minimum oil film thickness was straightforward and proceeded according to the model derived in Section 2. Mode 1 operation presented greater problems inasmuch as both upper and lower pads are in contact. The problem becomes that of calculating ho_1 and ho_2, the minimum oil film thicknesses under the upper and lower pads, in such a way that the loads they bear are equal to those calculated in the piston movement program.

Although the latter shows the piston-to-cylinder inclination to be near zero in Mode 1, it is necessary for the hydrodynamic calculation to postulate a finite value, however small, for this quantity. The calculation process thus begins by determining the minimum oil film thickness ho_1 for the upper pad corresponding to its known load W1 and a guessed value of piston-to-liner inclination. This inclination and the value of ho_1 is then used to calculate ho_2 geometrically and the hydrodynamic model invoked to calculate the corresponding value of W2. This estimate of W2 is now compared with the given value derived from the piston movement program, the comparison is used to suggest a new angle of inclination and the process repeated until W2 is equal to the given value.

The AEconoguide piston modelled in these calculations is based on the one illustrated in Fig. 2 and some of the results are plotted against crank angle in Fig. 6. Fig. 6A shows the loads carried by the upper and lower pads as determined by the piston movement program. Fig. 6B gives the corresponding piston-to-bore inclinations which largely determine the number of pads in contact at any particular crank-angle. Figs. 6C and 6D give the calculated film thicknesses under the upper and lower pads and the corresponding coefficients of friction respectively. Fig. 6E combines this data with piston velocity to show the instantaneous power loss through the operating cycle.

One of the first variables to be considered was that of aspect ratio, i.e. the ratio L/B. Fig. 7 shows calculated values of the average cyclic power loss and film thickness at 120° after tdc firing as a function of aspect ratio for 3 running conditions namely:

1600 rev/min. one third load
3000 rev/min. one third load
5000 rev/min. full load

All these calculations were performed for three individual pads on each side of the piston skirt, each of area 201.6 mm^2. The 120° point of the cycle was chosen for illustration of oil film thickness as being near to the position of greatest instantaneous power loss (Fig. 6E). Whilst the film thicknesses are indeed lower in some other parts of the cycle (Fig.6C), namely near the dead centres, the low forward velocities at such points means that they make only a small contribution to the integrated cyclic power loss.

One of the unknown quantities in setting out to perform the above calculations was the oil film thickness left on the cylinder bore. This was initially assumed to be 15μm, a figure chosen to reflect the film thickness in excess of the sum of the surface asperity heights needed to achieve the hydrodynamic lubrication evidently predominant with ordinary pistons (say circa 5μm) and the tendency for a "bow wave" of oil to pile up in front of the pads. Later distinction was made between the oil film thicknesses to be expected on the liner during the up and down strokes. On the up strokes the oil encountered by the pads below the lower reversal position of the oil control ring will be the oil left by the pads on the down stroke. Above this level it will be the oil film left by the oil control ring during its upward movement. On the down strokes, however, an increased supply of oil for the pads will have been provided by oil splash that occurred during the up stroke. Since therefore the oil film thickness available to the pads on the down stroke is evidently greater than the corresponding value of the up stroke, an alternative series of calculations were performed in which these 2 film thicknesses were taken as 30μm and 10μm respectively. The results for liner film thicknesses of 15μm and 30/10μm are plotted in Figs. 7A and 7B respectively.

In all the calculations the oil viscosity was assumed to be 16 mPs, corresponding to a 30 SAE grade oil at 90°C, approximately the mean temperature of the cylinder. The effect of total pad area is illustrated in Fig. 8 at engine running conditions of 1600 rev/min. one third load, 3000 rev/min. one third load and 5000 rev/min. full load. At each running condition the pad area is varied from a value well below that used in practice to one that would occupy most of the available skirt area. In each instance two sets of calculations are illustrated, for aspect ratios of 1.4 and 2.4.

(5) MEASURED PERFORMANCE IN OPERATING ENGINES

Measurements of the performance of AEconoguide pistons in operating engines, undertaken by both AE Developments and a number of potential customers, are described in Ref. 1, and cover the period up to the end of 1983. The situation in the Autumn of 1984 is summarised in Table 6. The precise improvement obtained is always a function of engine operation condition. Thus in Fig. 9 improvements measured by a potential customer vary as a function of bmep and engine speed from around 1.5 to 3.4% of fuel consumption, with almost all operating conditions in the band 2 - 3.4% improvement. So although the figures quoted in Table 6 are the best obtained under each of the operating conditions, the results of Fig. 9 show, by way of example, how such improvements were widely spread throughout the engine's operating range.

The accurate measurement of improvements in fuel consumption requires careful experimentation and it is difficult to ascribe a probable error. However, the four estimates of best improvement at part load given in Table 6 come from different testing stations and may be summarised as $(3.4 \pm 0.5)\%$ with 95% confidence.

(6) DISCUSSION

The calculations illustrated in Fig. 7A indicate that the power loss of AEconoguide pads of 201.6mm^2 area decreases with increasing aspect ratio L/B.

As would be expected, the power loss for any

given design of pad increases with speed and load. The film thickness under the pad shows a much smaller variation with operating condition, increasing with aspect ratio. Thus, from the point of view of both power loss and oil film thickness, a high aspect ratio is to be preferred, though in practice this would be limited by the effects of the piston geometry on the maximum extent of its contact with the bore in the circumferential direction. The results of Fig. 7B show a remarkably similar trend indicating the relative unimportance of the oil film thickness on the liner within the range considered. Since, however, assumed liner oil film thicknesses of 10μm on the up stroke and 30μm on the down stroke were considered the more realistic, these were used for the remainder of the calculations.

In Fig. 8 the hydrodynamic power loss is seen to fall rapidly with pad area, for both of the aspect ratios considered, namely 1.4 and 2.4. Calculated oil film thicknesses showed a similar trend, confirming the simplified calculations of Section 2 and underlining the dangers of reducing the pad area too much. As would be suggested by Fig. 7 the high ratio of 2.4 is seen to be preferable both on grounds of lower cyclic power loss and higher oil film thickness.

The oil film thicknesses under the conditions considered in Fig. 8 are summarised in Table 7. All of the film thicknesses increase monotonically with engine speed despite the higher load at 5000 rev/min.

In addition to the film thickness results, Table 7 includes the measured engine output power at each of the operating conditions. Taking as an example the results calculated for an engine speed of 3000 rev/min. and a pad aspect ratio of 1.4 (Fig. 8), the calculated viscous losses from 560 mm^2 pads on all four pistons amounts to 100W, corresponding to 0.65% of output power. As the pad area is decreased to 80mm^2 the total loss from 4 pistons falls to 58.4W, a reduction of 0.27% of output power. Although the total contact area for the piston with three 560 mm^2 pads on each side of the skirt is probably as great as that of a conventional piston, the detailed lubrication of the skirt is quite different, so that in no sense can the viscous losses for the piston be approximated to those of a conventional design. However, even so, it does seem that the calculated viscous power losses are too low. This is a problem encountered in other areas of hydrodynamics and here is probably due to the omission of surface roughness and boundary friction effects from the relatively simple model. For two real surfaces in relative motion, the finite surface roughness begins to influence the nature of the hydrodynamic lubrication as soon as the film thickness falls below about 3 times the sum of the asperity heights on the opposing surfaces (Ref. 8). As the surfaces approach closer, asperity interactions increase and the nature of lubrication gradually changes from hydrodynamic to boundary, with a further increase in the coefficient of friction. Including surface roughness effects (based on Ref. 8) in piston ring lubrication calculations at AE Developments has had the effect of increasing the calculated friction several times.

(7) CONCLUSIONS

Simultaneous solution of the pressure and temperature distributions throughout the film volume under a single pad operating at a constant inclination showed that the temperature distribution within the film was governed almost completely by the temperatures of the pad and the cylinder wall i.e. by direct transfer of heat across the film rather than generation of heat within it. This established that the AEconoguide pads did not work by causing local heating with a consequent reduction in viscosity. Multiple regression of the calculated hydrodynamic results for such pads led to the establishment of power formulae involving the pad dimensions, viscosity, velocity, load and inclination in which the powers differed significantly from simple theory (Tables 4 and 5). Combining the results of this regression analysis with calculations of piston angular movement (verified by measurements) allowed the calculation of oil film thickness and viscous loss for pistons with pads operating on a number of different engine cycles. The results of these calculations over complete engine cycles confirmed the trends established by simple theory. Thus high aspect ratio pads i.e. pads having a high ratio of circumferential length to axial height, were shown to have least viscous loss and greatest oil film thickness. However, in practice the extent to which such pads can be utilised will be limited by the maximum circumferential contact permitted by piston geometry, especially for the larger pad areas.

Alternative assumptions concerning the oil film thickness left on the liner in front of the pads (15μm under all conditions or 30μm on down strokes combined with 10μm on up strokes) were shown to result in very similar trends in the variation of power loss and oil film thickness with aspect ratio. Further calculations based upon the latter assumption and considering 2 values of aspect ratio, 2.4 and 1.4, showed the former to be preferable in terms both of lower viscous loss and higher oil film thickness. As predicted by the basic theory,the viscous power loss was seen to decrease with decreasing pad area down to the minimum considered value of 80mm^2 . Under these conditions the minimum cyclic oil film thicknesses might well be giving cause for concern, so practical values of pad area are significantly higher.

The calculated viscous losses were seen to increase with increasing combinations of speed and load, though no formal attempt was made to separate the two variables.

The calculated trends thus confirmed the operating principles of the AEconoguide design, but the absolute level of the calculated losses was probably too low, due, at least partly, to the omission of surface roughness and boundary friction effects from the relatively simple model. Fuel economy and power gains reported by customers have varied with operating conditions up to 2 - 3.8% and 1 - 3.8% respectively.

(8) REFERENCES

(1) RHODES, M. L. P. and PARKER, D. A. 'AEconoguide - The Low Friction Piston' SAE Paper 840181.

(2) HADDAD, S. D. 'The Origins of Noise and Vibration in Diesel Engines with Emphasis on Piston Slap' Ph.D. Thesis, University of Southampton, 1974.

(3) LAWS, A. M., PARKER, D. A. AND TURNER, B. 'Piston Movement in the Diesel Engine' CIMAC Paper No. 33, 1973.

(4) RHODES, M. L. P. and PARKER, D. A. 'The AEconoguide Low Friction Piston Skirt Design' Paper 30, AE Technical Symposium 1982.

(5) ETTLES, C. M. McC. Private Communication, 1982.

(6) CAMERON, A. 'Principles of Lubrication' Longmans 1962, p.116.

(7) PARKER, D. A., ADAMS, D. R. and GRAHAM, N. A. 'Some Contributions of Instrumentation to the Understanding and Prediction of Piston and Ring Behaviour' ISATA, Graz, Austria, September 1979.

(8) PATIR, N. and CHENG, H. S. 'Application of Average Flow Model to Lubrication Between Rough Sliding Surfaces' ASME J. Lub.Tech., Vol. 101, April 1979, pp 220 - 230.

TABLE 1 - DATA CHOSEN FOR INVESTIGATION OF TEMPERATURE DISTRIBUTION UNDER SINGLE PLAIN PAD

Independent Parameters	Dependent Parameters
B = 15 mm	Load W = 695.1 N
L = 15 mm	
h = 3.5 μm	Friction Force F = 1.88 N
Convergence $(h_1/h_o - 1) = \kappa = 2$	Coeff.of Friction μ = 0.00270
	Viscous power loss = 18.75 W
Flare, σ = 1.01	
U = 10 m/s	Conducted Heat Flux = 163.0 W
Pad Temp. = 140°C	
Wall Temp. = 100°C	
Average Viscosity, $\eta_{\frac{1}{2}}$ = 0.0066 Pa.s	
Reynolds Number, Re = 10.7	

TABLE 2 - CALCULATED FILM THICKNESSES FOR SINGLE PLAIN PAD - INPUT DATA AS IN TABLE 1. (NODES DEFINED IN FIGURE 4A)

	FILM THICKNESS MICRON								
	K=3	K=4	K=5	K=6	K=7	K=8	K=9	K=10	K=11
J = 11	14.030	13.155	12.280	11.405	10.530	9.655	8.780	7.905	7.030
J = 10	12.486	11.611	10.736	9.861	8.986	8.111	7.236	6.361	5.486
J = 9	11.383	10.508	9.633	8.758	7.883	7.008	6.133	5.258	4.383
J = 8	10.721	9.846	8.971	8.096	7.221	6.346	5.471	4.596	3.721
J = 7	10.500	9.625	8.750	7.875	7.000	6.125	5.250	4.375	3.500*
J = 6	10.721	9.846	8.971	8.096	7.221	6.346	5.471	4.596	3.721
J = 5	11.383	10.508	9.633	8.758	7.883	7.008	6.133	5.258	4.383
J = 4	12.486	11.611	10.736	9.861	8.986	8.111	7.236	6.361	5.486
J = 3	14.030	13.155	12.280	11.405	10.530	9.655	8.780	7.905	7.030

* Minimum

TABLE 3 - MINIMA AND RANGES OF THE VARIABLES FOR MULTIPLE REGRESSION

	Minimum	Range
Speed U,m/s	2	18
Length B, mm	2	18
L/B ratio	0.2	2.8
Convergence, α	0.01	8
Flare, δ	0.02	3
Wall Temp., °C	80	140
$T_{pad} - T_{wall}$ °C	5	95
Pressure, Pa	3×10^5	1×10^7

TABLE 4 - SUMMARY OF CURVE FITTING AND REGRESSION RESULTS FOR COEFFICIENT OF FRICTION

Conditions	Exponent of B	U	η	W	L	α	Result No.
Curve Fitting for Rectangular Pads	0.45	½ (Assumed)	½ (Assumed)	-½ (Assumed)	.05	-0.35	
Regression for Pads with Plane Convergent Wedge Films	0.416 ± 088	0.499 ±.054	0.507 ±.055	-0.586 ±.030	0.207 ±.064	-0.362 ±.019	(3)
Regression for Pads with Flared Convergent Wedge Films	0.581 ±.084	0.462 ±.046	0.495 ±.046	-0.543 ±.020	0.055 ±.055	-0.347 ±.019	(23)
Regression for Pads with Inlet and Outlet Ramps and Flared Convergent Wedge Films	0.569 ±.069	0.499 ±.061	0.378 ±.055	-0.544 ±.039	-0.003 ±.032	-0.326 -.267	(41)

TABLE 5 - SUMMARY OF CURVE FITTING AND REGRESSION RESULTS FOR MINIMUM FILM THICKNESS

Conditions	Exponent of B	U	η	W	L	κ	Result No.
Curve Fitting of Rectangular Pads							
$\kappa < 1.125$	0.55	$\frac{1}{2}$	$\frac{1}{2}$	$-\frac{1}{2}$	0.95	0.125	
$\kappa > 1.125$	0.55	$\frac{1}{2}$	$\frac{1}{2}$	$-\frac{1}{2}$	0.95	-0.20	
		←—	Assumed	—→			
Regression for Pads with Plane Convergent Wedge Films							
$\kappa < 1.125$	0.443	0.495	0.437	-0.473	0.935	0.396	4
	±.076	±.045	±.052	±.034	±.062	±.026	
$\kappa > 1.125$	0.662	0.532	0.502	-0.479	0.861	-0.216	5
	±.067	±.043	±.036	±.022	±.046	±.048	
Regression for Pads with Flared Convergent Wedge Films							
$\kappa < 1.125$	0.519	0.489	0.516	-0.495	1.090	0.436	24
	±.143	±.069	±.051	±.024	±.067	±.029	
$\kappa > 1.125$	0.495	0.493	0.529	-0.485	0.923	-0.168	25
	±.037	±.021	±.024	±.009	±.026	±.031	
Regression for Pads with Inlet and Outlet Ramps and Flared Convergent Wedge Films	0.364	0.467	0.508	-0.452	1.053	0.039	43
	±.040	±.035	±.032	±.023	±.048	±.031	

TABLE 7 - CALCULATED MINIMUM OIL FILM THICKNESS BETWEEN PADS AND CYLINDER BORE

Values at 120° ATDCF assuming oil film thicknesses in front of pads are 30μm and 10μm in downstrokes and upstrokes respectively.

ENGINE RUNNING CONDITION	ASPECT RATIO L/B	OIL FILM THICKNESS IN MICRONS	
		80mm² pad areas	560mm² pad areas
1600 rev/min	2.4	5.7	18.2
1/3 load (6.0kW)	1.4	4.6	15.4
3000 rev/min	2.4	5.9	18.9
1/3 load (15.3kW)	1.4	4.9	16.1
5000 rev/min	2.4	6.2	19.7
Full load (68.2kW)	1.4	5.1	16.8

TABLE 6 - AECONOGUIDE GAINS REPORTED BY 10 AE CUSTOMERS

	AECONOGUIDE SPECIFICATION		FUEL ECONOMY GAIN		MAXIMUM POWER GAIN	FRICTION REDUCTION
			Part Load	Full Load		
PETROL	No. of Attempt	Design Variant				
	1st	Early	3.4%	3.7%	1.8%	
	2nd	Early			1%	
	1st	Early				10/13% piston friction
	2nd	Current			2%	
	1st	Current	3.8%		1.5/3.8%	5% "friction power"
	1st	Current				5% "friction power"
DIESEL						
	1st	Early	3.5%	(2.5% (*3.7%	2.5%	
	1st	Early	3%	2.5%		
	1st	Current		2%		
	2nd	Early		2%		

* After further 160 h.

5 Petrol 1.0 - 4.1 litres
4 Diesel 2.5 - 7.9 litres
1 Single Cyl. Lab. Engine (Petrol)

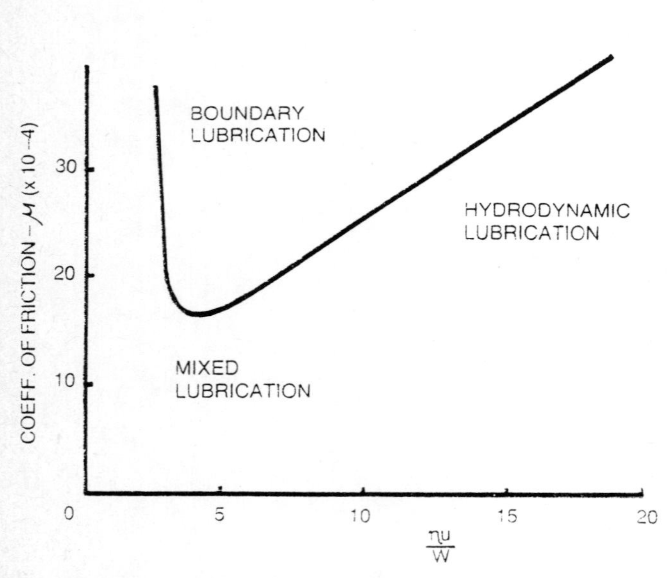

Fig 1 Stribeck curve for a journal bearing with single grade SAE 20 oil at 150°C

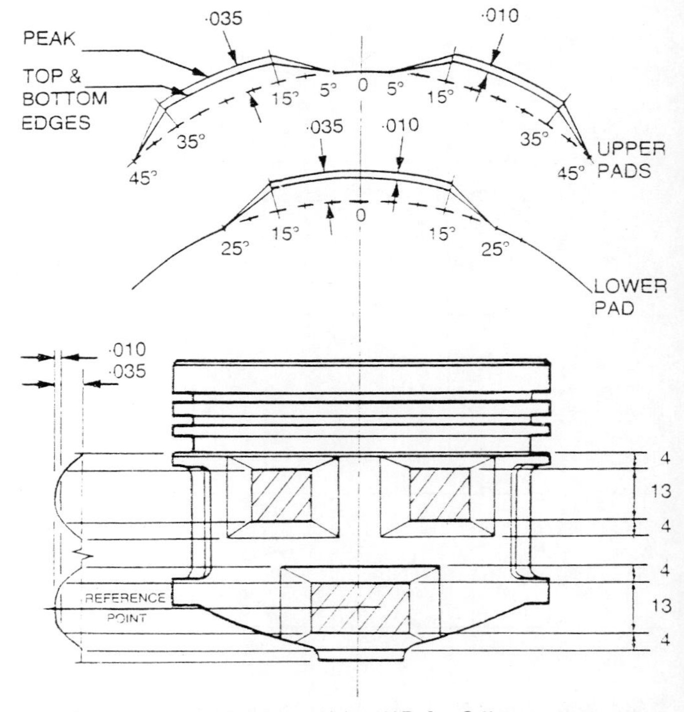

Fig 2 Æconoguide piston skirt IIID for 2 litre engine, ϕ 90. All dimensions in mm

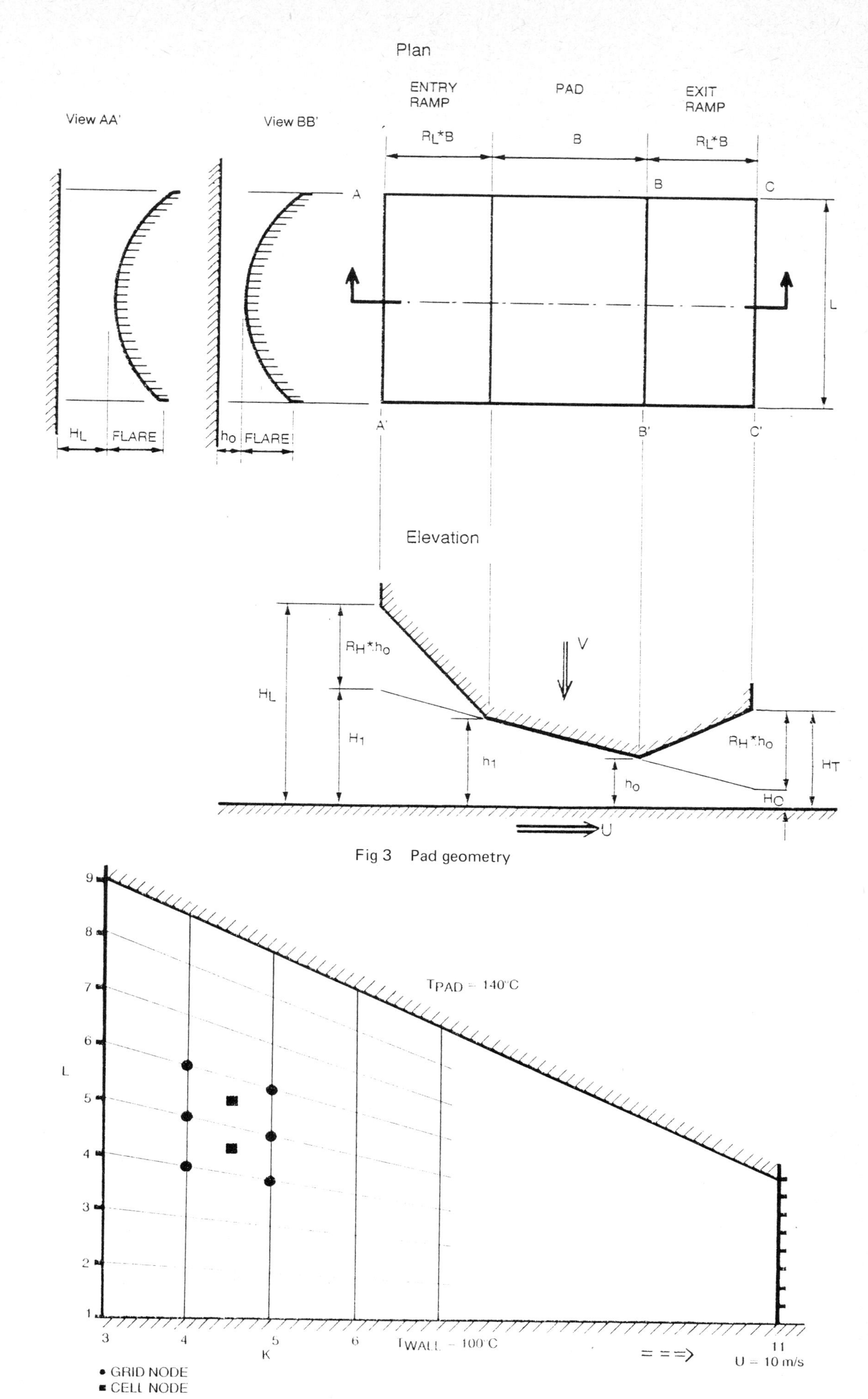

Fig 3 Pad geometry

Fig 4a Temperature distribution through film under single plain pad — subdivision of film volume

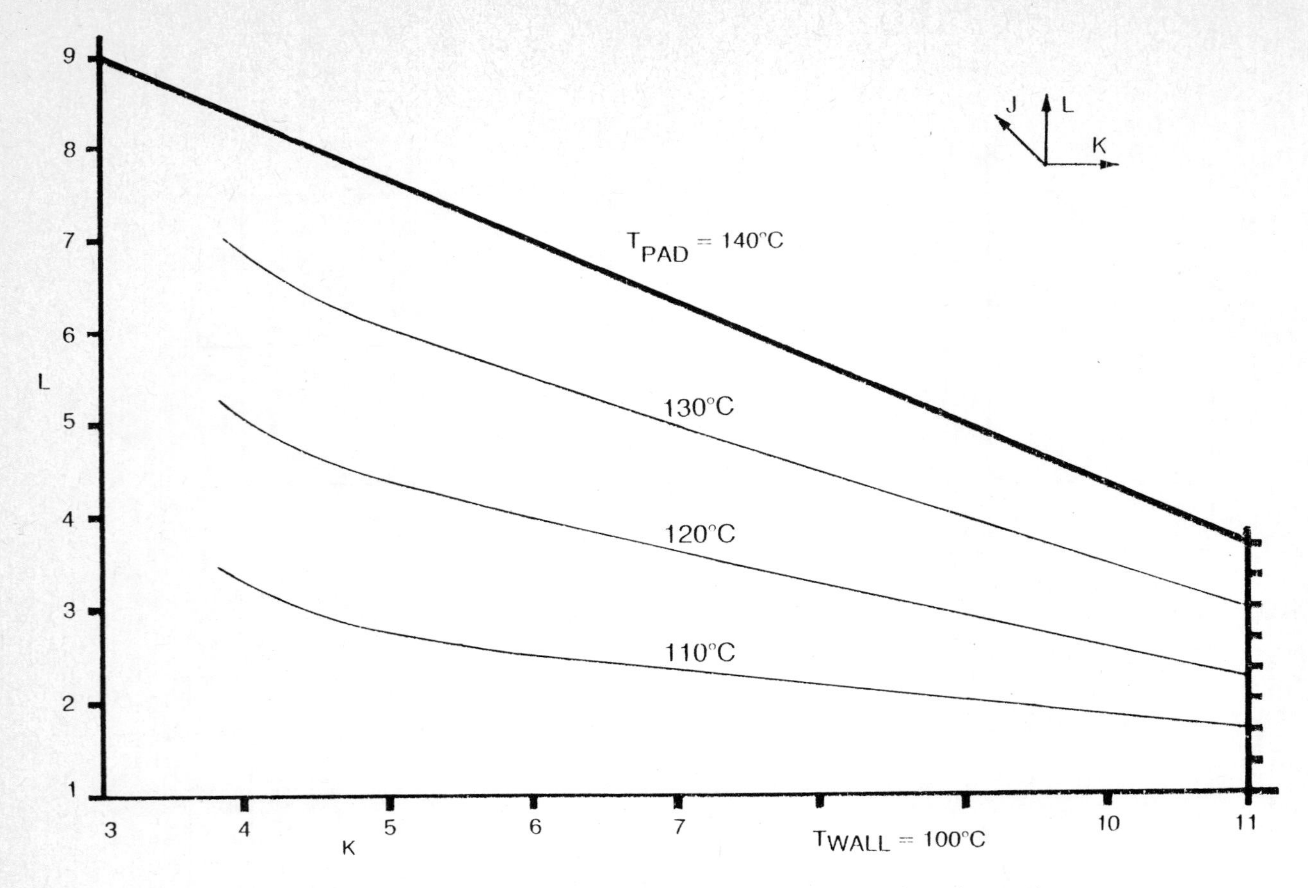

Fig 4b Temperature distribution through film under single plain pad — results for example of Section 2.3

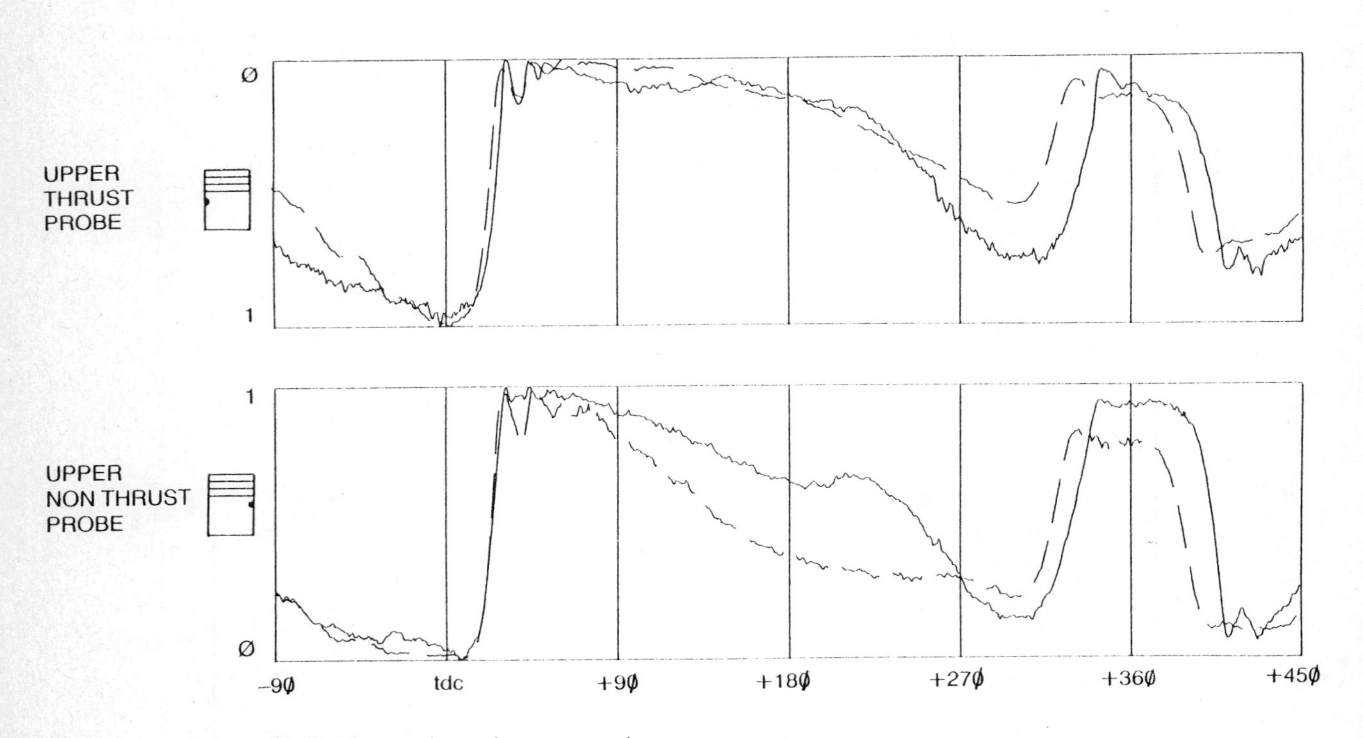

Fig 5 Comparison of transverse piston movements
——— Monometal piston with Æconoguide feature
– – – Standard monometal piston
2 litre gasoline engine one minute after cold start 1600 r/min, 1/3 load

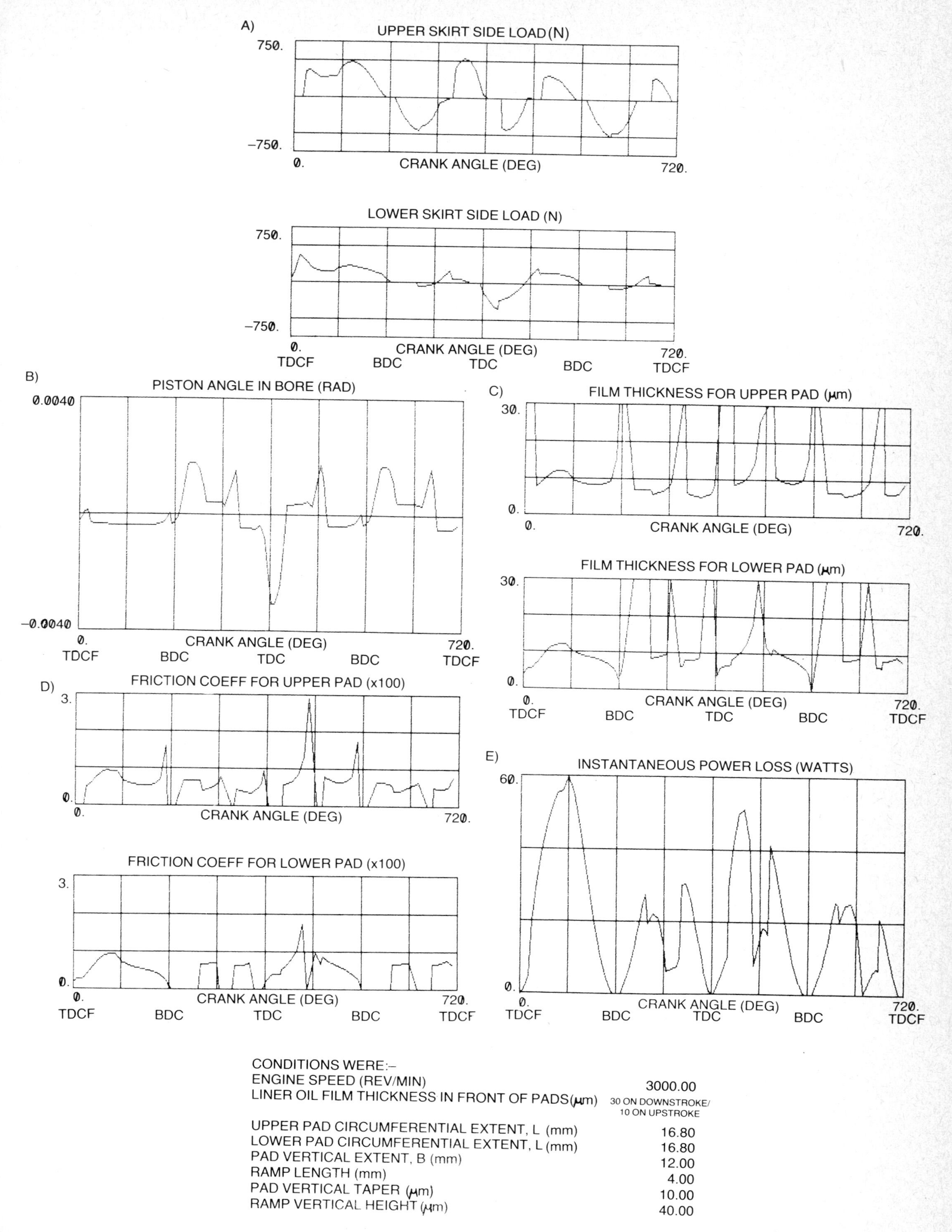

Fig 6 Æconoguide performance calculations for a 2 litre engine

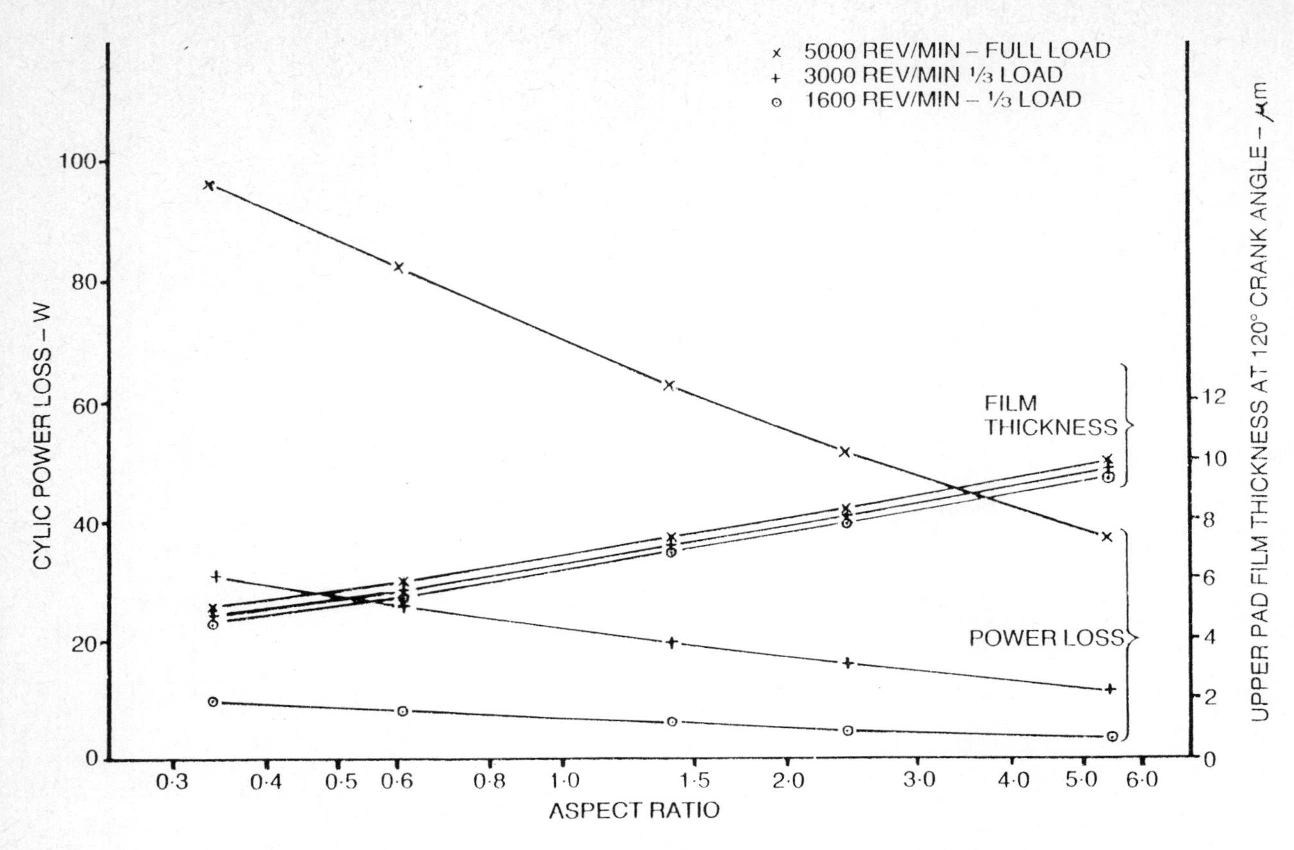

Fig 7a Æconoguide piston — calculated power loss and film thickness as a function of pad aspect ratio. Film thickness in front of pads — 15 μm on all strokes

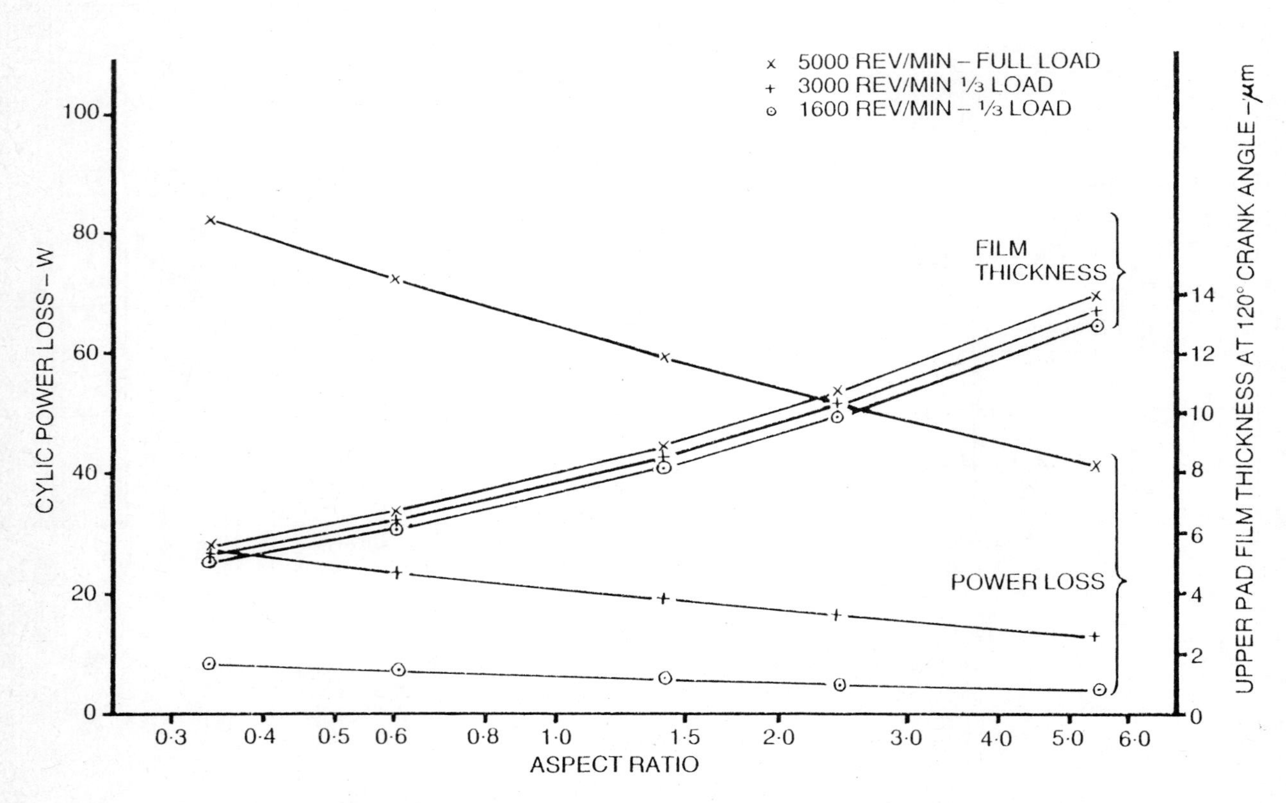

Fig 7b Æconoguide piston — calculated power loss and film thickness as a function of pad aspect ratio. Film thickness in front of pads —30 μm on downstrokes, 10 μm on upstrokes

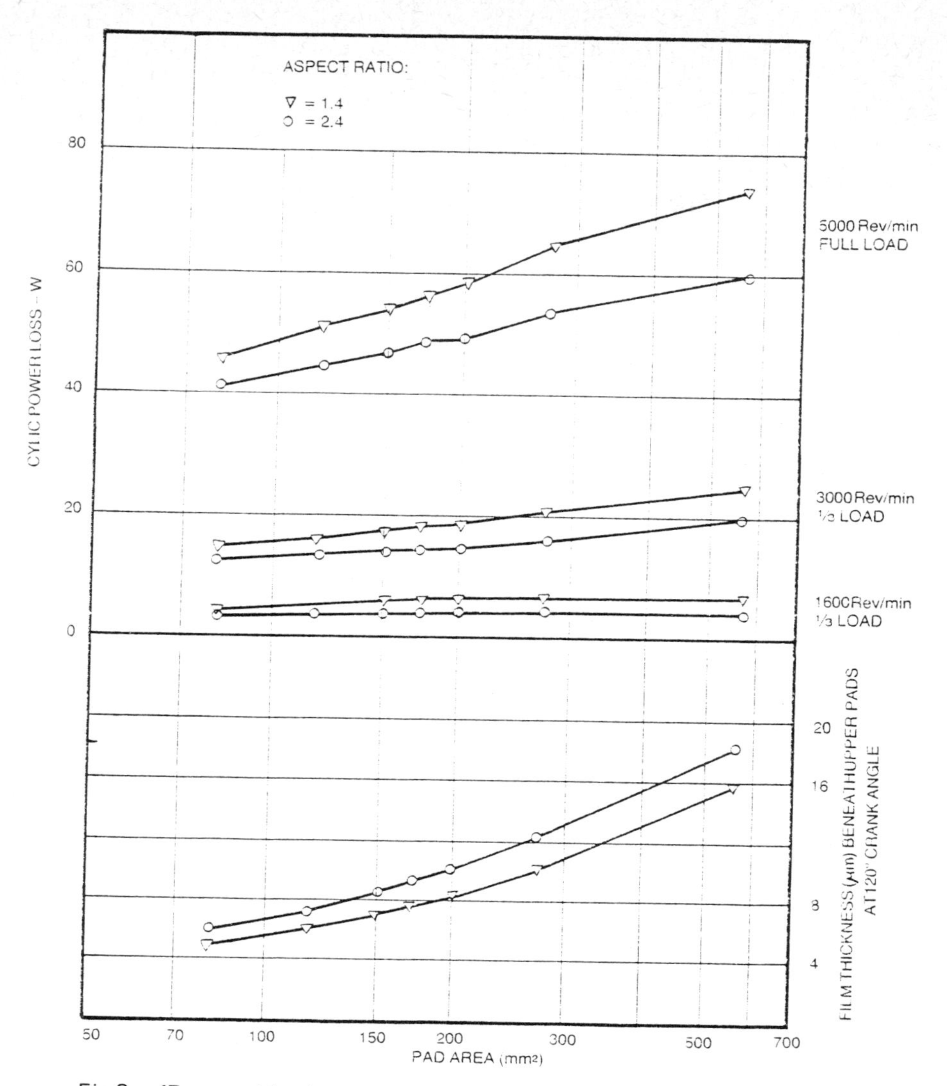

Fig 8 Æconoguide piston — calculated power loss and film thickness in front of pads as for Fig 7b

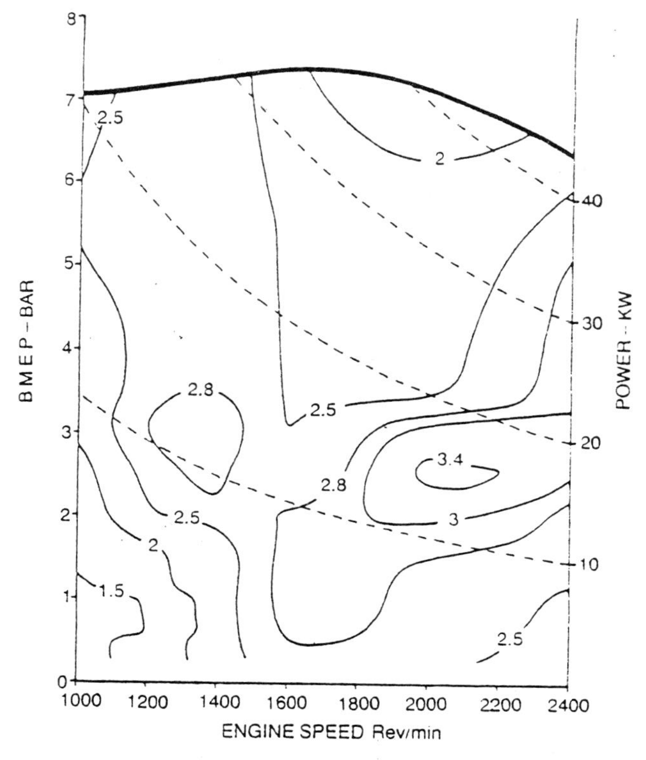

Fig 9 Percentage improvement in fuel consumption due to Æconoguide 3.5 litre diesel engine

C73/85

The influence of design parameters on engine friction

J D POHLMANN, Dr-Ing and H-A KUCK, Dr-Ing
Volkswagenwerk AG, Aggregate-Entwicklung, Wolfsburg, West Germany

SYNOPSIS

Based on the analysis of a large number of friction measurements with different piston engines a friction model was found which allows the calculation of total engine friction. In addition to that separate components were investigated to find their influence on friction. Thus, the model allows the **calculation** of separate components concerning friction.

The quality and potential of the model is demonstrated by a comparison of test bed results and calculation. Some examples are shown to give an idea of the influence of several parameters on engine friction, engine data as bearing dimensions, piston design, crankshaft drive, number of cylinders etc.

Introduction

The product of a technical development is always a relative optimum, in other word the best possible tradeoff between a wide range of possible solutions, costs and technical feasibility. The actual series version is greatly influenced by the financial outlay for design, testing and production in particular as well as by the time frame available. Thus, it is worthwile making every effort to reduce, for example, the extent of testing and the expense it involves. A familiar method of achieving this goal is to make use of simulation measures involving advanced computer systems, which make it possible in the pre-development stage to optimize various parameters by forming computational models. Such work need not necessarily be focused on the basic design feasibility, but rather it is possible to calculate the potential of a concept (possibly the optimum) and then approximate it to a greater or lesser degree in practice. A computational simulation thus makes it possible to perform any number of parameter studies and to determine individual influences on a largely independent basis. If real testing were to be employed, this would involve a huge outlay in terms of time and material (1).

The value of such calculation is largely dependent, however, on the quality of the computational model. The accuracy, for example, with which the physical properties of an engine can be described will determine the confidence level of the results. This does, however, require knowledge of a wide range of parameters which can often only be obtained by employing sophisticated measurement methods. The object of this paper is thus to draw a comparison between a relatively simple simulation program for calculation of engine friction and selected, more sophisticated computational models.

1) Computational Model for Engine Friction

1.1 General

It would obviously be desirable to have an extremely simple, physically correct simulation model, which would enable engine friction to be calculated in advance with a minimum of known parameters. This is, however, scarcely feasible given the large number of factors influencing the functioning of a combustion engine.

Some examples are quoted below.

a) geometric dimensions of components causing friction

b) structural design of components causing friction

c) type and number of components causing friction

d) physical properties of components causing friction and lubricants used

e) component temperatures

f) type of engine etc.

Each item in the list can be broken down into further detailed data in order to obtain all the relevant parameters. It is thus no surprise that information is rarely given in pertinent literature regarding calculation of overall engine friction. The wide range of publications available almost always only make it possible to derive the effect of quite specific parameters on the friction of a particular assembly. Each computer program is extremely detailed and requires numerous input variables which can only be obtained employing sophisticated measurement methods.

1.2 Procedure

Here a different approach was adopted. A purely empirical friction program was produced on the basis of a large number of measurements taken on engines with varying design principles. This program is capable of calculating friction as a function of load, speed and oil temperature for the engines under consideration. The next step was to split up the model into modules in order to be able to calculate the effect of individual assemblies on overall friction. These correlations are likewise purely empirical and describe experimental data.

However, in order to be able to use the computer program to calculate in advance the frictional behaviour of other engines as well, similitude ratios were derived as far as possible for each module, thus making it possible to transfer results to identical assemblies with other geometric dimensions whilst taking into consideration the mechanical loading of the components causing friction. (2)

Finally, each module can be configured for the corresponding assembly by way of a sophisticated special computer program, with all the consequences this entails in terms of the input data required. Once such a program package has been established for each module, a computer program is available which can determine every factor influencing friction in isolation. In many cases the data required for the marginal conditions involved with particular problems are not available, with the result that recourse must be had to a less sophisticated program version.

1.3 The Model

As already indicated in Section 1.2, success was achieved in breaking down various engines into their individual friction components using, in some cases, extremely complex measurements. (3,4,5)
This involved the following components:

a) the cylinder head (valve gear)

b) crankshaft group

c) piston group broken down into

   1) piston skirt
   2) compression rings
   3) oil ring

d) bearings

e) auxiliary assemblies.

Each of the above-mentioned items represents a module of the model and describes the friction conditions in the computer program with semi-empirical correlations. This is explained in the following taking bearing friction as an example. (Fig. 1)

The measurements taken resulted in the illustrated correlations for the main bearings of an engine. As is known, bearing friction increases with speed in particular, whereas the load on the bearing has less of an influence presupposing hydrodynamic lubrication.

This is especially true of bearings for combustion engines, where the load, i.e. the cylinder internal pressure during combustion, only takes up a limited period of time on account of the cyclic loading over the course of a working cycle. Assuming that bearings featuring the same design principle are similar in terms of friction, a relationship between bearing width and bearing diameter given identical mechanical loading can be derived from the theory of hydrodynamic lubricant films.

Presupposing an adequate number of measured bearings of different designs, it is possible in this way to achieve a semi-empirical calculation of bearing friction. Semi-empirical correlations can also be established in the same manner for the piston group for example, so as to determine the influence of the number of rings, the piston-skirt area and the bore diameter etc.

The computer program thus makes it possible to calculate friction whilst taking into account the following parameters:

a) stroke and bore

b) number of cylinders

c) number of bearings, geometric dimensions of bearings

d) crankshaft group kinematics (horizontal displacement, axial displacements)

e) piston geometry, ring configuration

f) load

g) speed

h) oil temperature

i) various auxiliary assemblies such as oil pump, water pump, injection pump etc.

Several basic correlations are illustrated in the following on the basis of appropriate diagrams.

2. Calculations Performed

2.1 Comparison of Various Models

As can be seen from the illustration 1, the friction mean effective pressure of bearings in combustion engines which are subjected to nonsteady state loading can be reproduced with good approximation by way of a polynomial expression (solid line). The model used here is of the form described in the following equation

$$P_{mr} = a + bn + cn^2 + \ldots \quad E\ (1).$$

Experience has shown that the polynomial expression should be terminated with the power of 2 since undesired inaccuracy occurs with a polynomial of a higher degree due to vibration between the interpolation points. Restriction to a fitted polynomial of the 2nd degree on the basis of a wide range of measurement data produced measurement results with a high degree of correlation.

Equation 1 does, however, only take account of the speed dependence of the friction mean pressure. Further parameters, such as load and engine-oil temperature, are likewise contained in the computer program in the form of fitted polynomials. These fitted polynomials determine the coefficients a, b and c in equation 1 and thus make it possible to establish the friction mean pressure as a function of speed, load and oil temperature. Application of the laws of similitude enables the results obtained to be transferred to sliding bearings of other geometric dimensions.

$$P_{mr} = \frac{BD^3}{B_o D_o{}^3} \bullet (3{,}04 \cdot 10^{-2} + 3{,}02 \cdot 10^{-5} \cdot n + 4{,}3 \cdot 10^{-9} \cdot n^2) \quad (E\ 1a)$$

for bmep = 2 bar

ϑ_{oil} = 80 °C

The friction loss of a sliding bearing is proportional to the bearing width and the cube of the diameter.

$$N \sim B \bullet D^3 \quad (E\ 2)$$

Taking into account the similitudes indicated in (equation 2) a calculation of other basically identical bearings is also feasible.

The data input is thus limited to parameters such as load, speed, oil temperature and bearing dimensions. The figure also illustrates - in the form of triangles - the results of far more complex calculations starting from the situation outlined above. These are based on calculations of journal movement in friction bearings subjected to non-steady state loading in accordance with the theory of hydrodynamic lubricant films. This theory is known from pertinent literature (6,7,8) and need not be described in greater detail here.

A comparison of the input data required for reliable calculation is, however, interesting (Table 1).

Table 1

Input Data for Calculating Displacement Path of Friction Bearings Subject to Dynamic Loading

1) all masses

   piston, piston pin, connecting rod, piston rings, counterweights etc.

2) complete geometry of crankshaft and bearings

3) bearing clearance

4) viscosity (here it is to be noted that the bearing temperatures are not constant over the periphery)

5) cylinder pressure profiles as a function of speed and possibly also load.

Even allowing for the illustrated wide range of data to be input and which can thus be taken as being known, a total accuracy of the computational results is not guaranteed, with the result that, as the illustration shows, these friction mean pressures - calculated using sophisticated methods - can also be reproduced with a high degree of accuracy by the fitted polynomial. Similarly, the friction of the piston group can also be established using simple, empirical or sophisticated models. Reference is again made here to the extremely comprehensive data record for calculating piston friction in Table 2.

Table 2

Input Data for Piston Friction Program

1) all relevant masses

   piston, piston pin, piston rings, connecting rod

2) precise geometry of piston and rings

3) oil availability at friction locations and oil viscosity

4) clearance

5) description of cylinder-block deformation under the effects of pressure and temperature

6) precise description of piston movement in cylinder liner

7) cylinder pressure profiles as a as a function of speed and load.

In this respect the shearing forces in the load-bearing oil film are calculated assuming hydrodynamic lubrication conditions. Influence is exerted, for example, by temperature and load-induced deformation of pistons and engine power section, as well as piston movement and oil supply.

As was the case when calculating sliding-bearing friction, the major problem when attempting to establish piston friction is a lack of precise input data for engines which do not yet exist.

A simplified computational model,which only requires a limited number of input variables, thus appears both meaningful and justified for basic considerations, for example in the pre-development phase, particularly if the confidence range - which can be checked by way of calculation and/or measurement - is known. The minimal information level required for a unit, which has not yet been constructed, is given in Table 3.

Table 3

Input Data for Simplified Empirical Program (Total Friction)

1) stroke, bore, number of bearings and dimensions

2) oil temperature

3) load, speed

Thanks to the modular structure of the simulation program (Section 1.3) a more sophisticated calculation can however be performed at any time if the necessary data are available.

2.2 Influence of Design Parameters on Individual Friction Components (Fig. 1)

The illustration shows the friction mean pressure profile for the main bearings of a combustion engine. The shifts shown are the result of varying the bearing diameter by +/- 10 %. Width variation has been effected in the figure 2 for the same bearings. It clearly shows the tendency indicated by equation 2, namely that a change in diameter has a more pronounced effect. This is further reinforced in that the load-bearing capacity of the bearing increases in superproportional fashion with increasing bearing width, i.e. the bearing width increases in an underproportional manner given the same load-bearing capacity.

A further parameter, the variation of which is to be illustrated here by way of example, is the stroke-bore ratio s/D (Fig. 3). Given the marginal condition that the engine displacement remains constant, the effect becomes more pronounced with increasing speed. This can be plausibly explained by the change in piston speed due to the differing strokes. Referenced to identical mean piston speeds the differences are far less marked since the change in bore is then the only parameter to have an effect.

As regards the overall concept of an engine, which must satisfy certain specifications as regards consumption, output and smooth running, for example, questions relating to the number of main bearings or the number of cylinders are of significance at the development stage. The number of cylinders, in particular, influences the length of the engine and thus the possiblility of fitting it in the vehicle in question. Differences are, however, also to be expected as regards friction, and thus fuel consumption.

Fig. 4 illustrates the influence of the number of bearings on the friction of a 4-cylinder engine. Reducing the number of bearings decreases the friction mean pressure. The extent to which this measure has an effect depends, however, on the selected marginal conditions. A significant reduction can be achieved if, for example for reasons of maintaining identical parts for engines produced on a large scale, the number of bearings can be reduced without subjecting the bearings to excessive loading. If, on the other hand, the bearing dimensions must be adapted, the difference decreases considerably with the same diameter-width ratio and identical specific load and both solutions assume a virtually equal ranking as regards friction (Fig. 5).

The influence of the number of cylinders on the overall friction of an engine with constant displacement and constant stroke-bore ratio is illustrated in Fig. 6. The only difference is in the low-speed range (urban traffic). This difference should, however, be negligibly small taking into account, for example, smooth engine running. (1)

The breakdown of the overall friction of a 6-cylinder in-line engine is shown in Fig. 7. It can be seen that piston friction, bearing friction and also several auxiliary assemblies are responsible for a large proportion of overall engine losses. A justification is also provided here for the worldwide endeavours to precisely calculate in particular bearing and piston friction in advance. A relatively simple method well suited to indicating trends and which can also be used for the entire engine was presented here.

3) Summary

A comparison is drawn between a relatively simple simulation program for calculation of engine friction and selected, more sophisticated computational models. The program allows to calculate friction mean effective pressure as a function of load, speed, oil temperature and given design parameters.

Only a limited number of input data, which normally is given by design, is needed.

The comparison shows the potential and the quality of the model to predict tendencies concerning friction of different engine design parameters.

By this, the simulation program is a comfortable instrument of streamlining the scope of practical testing and thus to give in an effective format by way of preselection and preoptimization. It will not be possible to replace practical testing completely.

4) **References**

1) Pohlmann,J. D.., Kuck, H.-A.

Analysis and Computers Simulation of Design and Combustion Parameters Influencing the Fuel Consumption of Piston Engines

SAE 840180

2) Eberan v. Eberhorst, R.

Der Einfluß des Hub/Bohrungsverhältnisses auf die Triebwerksbeanspruchung VDI-Z, 99, 1957, No. 27

3) Hohenberg, G.

Reibungsanalyse 6-Zyl.-Motor 1981/82 (intern)

4) Szengel, R.

Motorreiung-Untersuchung des Reiberhaltens der Kolbengruppe Ifko 1983 (intern).

5) Oetting, H. Schwarze, G.
Ebbinghaus, W.

Friction losses in production gasoline engines and the effect of design and operating parameters

Las Vegas 1982

6) Groth, K., Pohlmann, J. D.

Bahnberechnung der Wellenverlagerung bei Kolbenmaschinen.

Ölhydraulik und Pneumatik

1985 , 17, No. 7

7) Butenschön, H.-J.

Das hydrodynamische, zylindrische Gleitlager endlicher Breite unter instationärer Belastung.

Diss. TU Karlsruhe 1976

8) Möhlenkamp, H.

Beitrag zur Berechnung der Lagerbelastungen und Verlagerungsbahnen schnellaufender Hochleistungsdieselmotoren bei verschiedenen Lastpunkten, insbesondere bei Drehzahldrückungen.

Diss. TU Hannover, 1974.

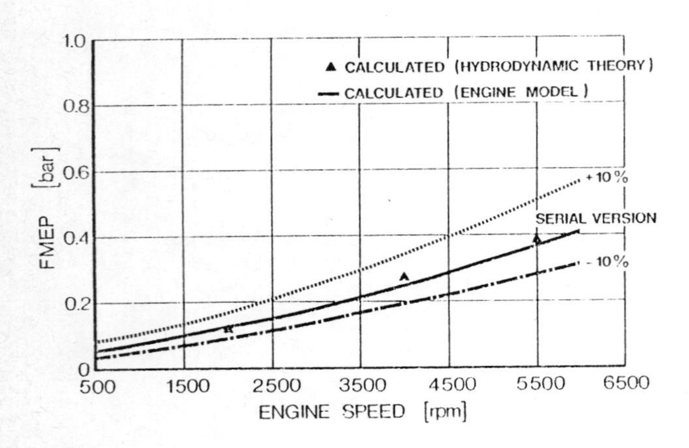

Fig 1 Friction of main bearings: variation of bearings diameter

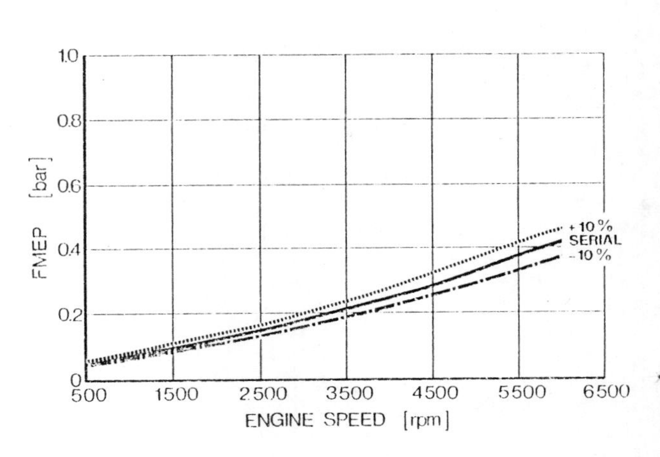

Fig 2 Friction of main bearings: variation of bearings width

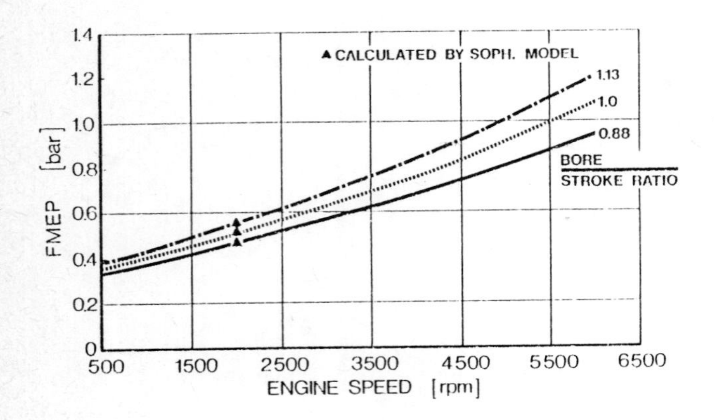

Fig 3 Friction of piston group: variation of stroke/bore ratio

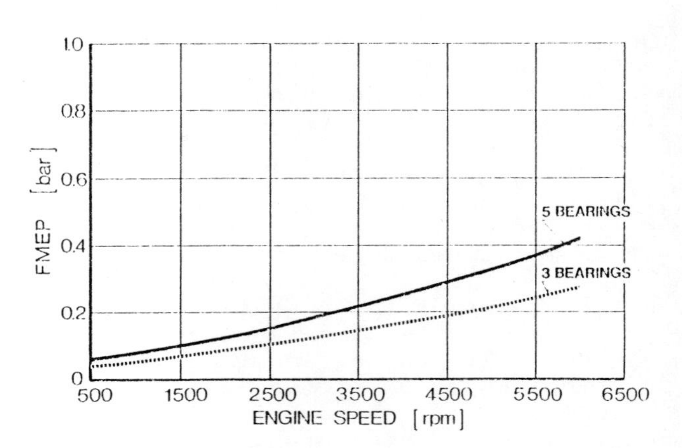

Fig 4 Friction of main bearings: variation of number of bearings — identical dimensions

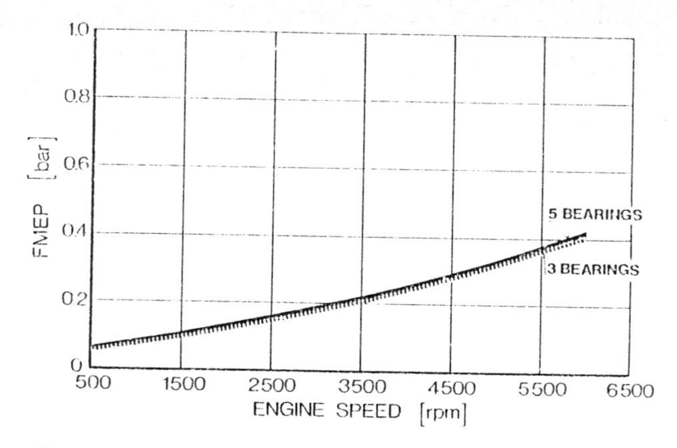

Fig 5 Friction of main bearings: variation of number of bearings — adapted dimensions

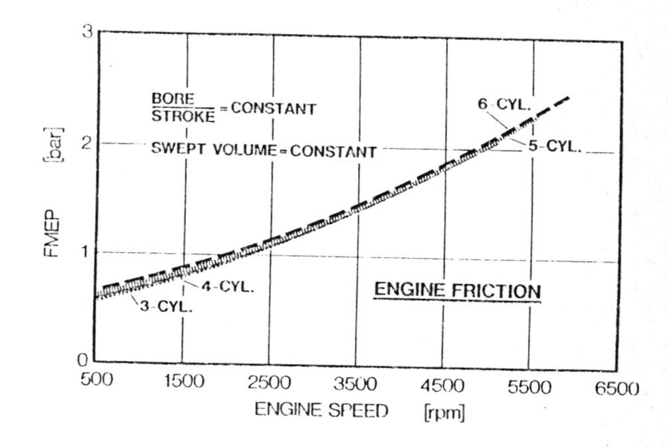

Fig 6 Engine friction: influence of number of cylinders

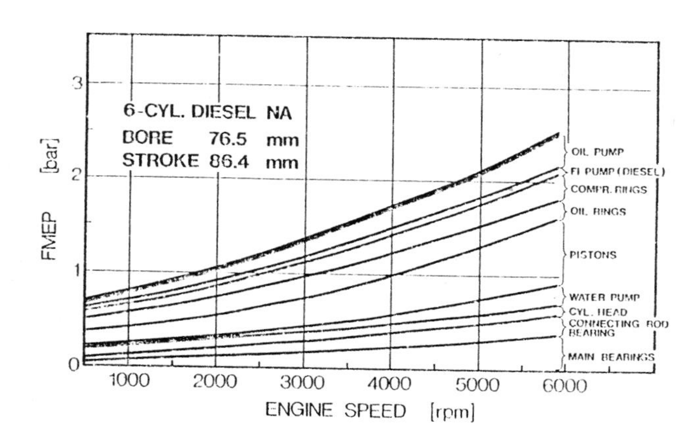

Fig 7 Overall engine friction: breakdown of overall losses

Potential fuel savings by use of low-friction engine and transmission lubricants

W BARTZ, Dr-Ing
Technische Akademie Esslingen, D-7302 Ostfildern 2, West Germany

ABSTRACT Applying lubricants is one of the most important measures to reduce friction and wear. By improving the mechanical efficiency of engines and gears by reducing the frictional losses in the mixed film as well as in the fluid film lubrication regime the reducing of the fuel consumption of the engine or of the energy needed to operate the cars is possible in principle.

Using the efficiency analysis of engines and gears an estimation of the theoretically possible maximum consumption reduction by lubricating measures is performed, evaluating the proportionate effects of lower viscosity and using friction reducing additives. The estimation is done on the basis of the given relationship between friction and fuel consumption. Obviously the possible consumption reductions at partial loads and lower temperatures are higher than at full load and operating temperature. By this fact the relevant consequences regarding the possible total energy reductions under political economy aspects are given. It will be shown that these states of operation, resulting in the highest reductions of fuel, only cover a relative small range of the total operating times of vehicles. Therefore the possible reductions are more limited than normally is assumed using especially unfortunate conditions as a basis for estimating the effects by lubricant-based measures. The results of this analysis leading to realistic full reductions are compared with published data, which will be critically evaluated.

1 INTRODUCTION

The necessity to reduce energy consumption will be the most important aim of the rest of this century. There is no doubt that the fuel economy of passenger cars has to contribute to this objective.

Of course this target cannot be reached only by tribological measures including using friction-reducing lubricants. Design modifications to the engine and the vehicle, weight reducing measures as well as improving the combustion process including the growing use of diesel engines and in addition speed and driving limitations will exert remarkable influences.

Though this is well known sometimes we are surprised by publications promising fuel consumption reductions of 10, 20 or 30 %. But the engineer should become extremely suspicious if he needs efficiency values of 100 % and more in order to achieve these targets. Therefore it will be the aim of this study, by analysing the efficiencies of engines and gears and the influences on them, as well as the friction modes in these units, to estimate the order of magnitude of the maximum possible fuel consumption reductions by lubricant changes.

These results will be compared with test results regarding fuel economy improvements by lubricants published in the literature.

The following considerations are related to the passenger car population, the types of vehicles and the driving behaviour in Germany.

2 FUEL ECONOMY IMPROVEMENTS PUBLISHED IN LITERATURE

In a recent comprehensive analysis an evaluation of published test data regarding fuel economy improvements using other lubricants in engines and gears has been performed (1). These results completed by newer investigations are summarized in Table 1 (2-15).

Table 1: Published test data regarding fuel consumption reduction by changes to engine and gear oils

	Fuel Consumption Reduction By		
	Lower Viscosity	Friction Modifier	Synthetic Lubricants
Engine Oils	0 - 5 %	0 - 9 %	0.8 - 4.2 %
Gear Oils	0.2-5.0 %	?	- 2.2 %
Engine and Gear Oils	1.1	-	11.0 %

It may be recognized that by optimum combinations of engine and gear oils improvements up to 11.0 % have been measured, depending on test conditions and the basis of comparison.

3 RELATIONSHIP BETWEEN FRICTION AND FUEL CONSUMPTION

3.1 Energy distribution in vehicles

A simplified representation of the energy distribution in a passenger car for a specific driving cycle is shown in Fig. 1 (16). At the output of the engine in this example only 15 % so-called useful power from the total input energy (energy capacity of the fuel) are available, which is reduced by further losses, e.g. in the gears, to not more than 12 % which will be available at the wheels to drive the car. Of course another distribution will be obtained for other driving and operating conditions, but the order of magnitude of the single components of power will be maintained.

There is no doubt that by tribological measures only the mechanical efficiency can be improved. Simplified it follows from this fact that by changes to gear and engine oils only the friction losses in these units can be influenced. For the example shown in Fig. 1, 7 % losses in the engine and 3 % losses in the gears may be effected. This will result in a fuel consumption reduction of 10 % if the friction losses in the engine and the gears were reduced to ZERO; truly an utopian imagination.

Of course there may exist less or more favourable conditions with higher or lower mechanical losses which can be influenced by changes to lubricants resulting in higher or lower fuel economy improvements.

3.2 Efficiency and fuel consumption

The total efficiency of a vehicle is given by the product of the individual efficiencies of the units in the power line. The following equation express this relationship:

$$\eta_{t_v} = \eta_{t_e} \cdot \eta_{t_g} \cdot \eta_{t_o}$$

with

η_{t_e} the total engine efficiency

η_{t_g} the total gear efficiency and

η_{t_o} the total efficiency of other units.

3.2.1 Engine

The total efficiency of an engine is defined by the product of thermal efficiency η_{th} and mechanical efficiency η_m. The mechanical efficiency is the ratio of useful power to indicated power.

$$\eta_{t_e} = \eta_{th} \cdot \eta_m$$

Values of the total efficiency are between 20 and 40 %. Depending on engine type the mechanical efficiency is between 78 and 92 %. This means that between 8 to 22 % of the indicated power is lost in mechanical losses.

Because reduced friction losses at a given condition by increasing the total efficiency will directly reduce the fuel consumption, the theoretically possible reductions will be estimated based on the minimum and maximum friction losses mentioned above. The result is shown in Fig. 2 revealing the percentage reduction of the friction losses. It is seen that fuel consumption reductions between about 5 % and 28 % (depending on the mechanical efficiency) are possible if mechanical friction could be eliminated. As already mentioned this is impossible but it seems to be realistic to reduce the friction losses by tribological measures up to 30 %. On this basis fuel consumption reductions of 11 % at unfavourable engine conditions and of 1.5 % at favourable engine conditions can be expected.

3.2.2 Gears

In this discussion only spur gears will be considered. The total gear efficiency is defined by the product of gear box efficiency η_g and the final drive differential gear efficiency η_a.

$$\eta_{t_g} = \eta_g \cdot \eta_a$$

The losses of a spur gear pair are the sum of the idling power, the tooth power losses under load and the bearing power losses under load. The efficiency of gears in the power train can be between 0.98 and 0.99. In addition there are 1 to 2 % losses by idling gears, bearing and seal friction and oil splashing. There shaft gear efficiency values of 0.95-0.96 for the lower gears and of 0.98 % for the direct gear can be assumed. Regarding the differential gear a lower efficiency between 0.94-0.96 has to be taken into account.

Obviously the mechanical efficiency of toothed-wheel gears is much higher than that of the engine. This fact will result in much lower potential fuel consumption reductions.

In Fig. 3 the range of possible fuel consumption reduction by improving the gear efficiency at unfavourable and at favourable conditions is shown. Provided a friction reduction of 30 % is realistic, fuel consumption reductions between 1.1 % and 3.5 % can be expected.

Of course the limiting values regarding fuel economy improvement changing engine and gear oils mentioned above are valid only for some vehicles and driving conditions. More detailed discussion is required on the scope for improvement which exists in average practical cases.

4 ESTIMATION OF FUEL CONSUMPTION REDUCTION BY MODIFIED LUBRICATING OILS

4.1 Basic relationships

It may be assumed that in engines about 2/3 of the friction losses are caused by fluid film friction and 1/3 by mixed or boundary film friction (8, 19). In the range of fluid film lubrication, the viscosity and in the range of mixed or boundary film conditions, the friction modifier, are the important influencing factors.

Simplified this means that the potential fuel economy improvements are reduced if only one of the lubricant-related measures is realized. In Fig. 4 the friction is shown schematically depending on speed. With increasing viscosities, the speed at which mixed film lubrication (characterized by wear) changes to fluid film lubrication (characterized by no wear) is shifted to lower values - an advantageous effect.

At higher speeds the friction increases considerably, however, a disadvantageous effect. At lower viscosities this friction increase will be less pronounced but higher speeds are needed for the transition between mixed film and fluid film lubrication.

The viscosity-temperature behaviour of oils is also of interest. With a high viscosity index the viscosity in the range of fluid film lubrication will decrease less with increasing speed due to friction heat, so that friction will increase much more than for oils with low values of the viscosity index. From this point of view a high viscosity index is not desirable in all cases. But most of the high viscosity index oils are non-newtonian oils characterized by decreasing apparent viscosities with increasing speeds or shear rates. By this effect the smaller fall in viscosity due to temperature compared to newtonian oils is balanced. These relationships are shown schematically in Fig. 5. These considerations result in the statement that multigrade oils having rather pronounced non-newtonian flow characteristics are desirable.

It follows from these considerations that in the range of fluid film lubrication the fuel consumption increases with increasing viscosity and that the curve for newtonian fluid lies below the curve for non-newtonian oils and develops with a lower shape. This picture is obtained if the low shear rate viscosities of the non-newtonian oils are plotted as shown in Fig. 6 schematically. If the viscosities of the non--newtonian oils are measured at the effective shear rate the same effective viscosity compared to a newtonian oil would be obtained resulting in the same fuel consumption. Due to different viscosity-temperature behaviour the shape of the curve, representing the non-newtonian oil is lower. These considerations have been confirmed experimentally (20), resulting in the statement that the effective viscosity is responsible for fluid film friction and fuel consumption.

It could be shown too that with increasing engine load the viscosity influence on fuel consumption decreases, obviously the result of increasing mixed film and boundary friction with increasing load. In the mixed film and boundary friction regime friction can be reduced by a so-called friction modifier. This effect is shown in Fig. 7 schematically. Such an additive is effective only if at certain conditions the viscosity is too low to maintain fluid film lubrication. In other words the effectiveness of such additives will decrease with increasing viscosity. It should be noticed that the ratio between mixed film and fluid film lubrication is important, having a certain value for a specific engine running at specific conditions. Trying to reduce the friction in the fluid film lubrication regime by reducing the viscosity it must be noted that the mixed film lubrication regime will be increased simultaneously. There will exist a critical minimum viscosity below which there will be an increase of friction again. This relationship can be explained schematically by Fig. 8. With a given oil the friction losses will decrease with increasing temperature due to decreasing viscosity. If the viscosity falls below a certain critical value a transition from fluid film lubrication to mixed film lubrication occurs resulting in increasing friction. This transition point is shifted to higher temperatures with higher viscosities. These considerations could be confirmed experimentally too (21) resulting in critical viscosities characterizing the transition between fluid film and boundary friction of 3.5 to 4.5 mms^{-1}.

Taking into account all effects which have been discussed a multigrade oil which should have a high viscosity index in order to cover the low temperature requirements should possess the following properties in order to contribute to fuel economy improvement by lower friction, avoiding simultaneously higher wear in the mixed film regime:

- Rather low viscosity at low shear rates
- Pronounced non-newtonian flow properties
- Incorporated friction modifiers.

4.2 Influence of viscosity reduction on fuel consumption reduction

4.2.1 Engine oils

In the fluid film lubrication regime friction depends on viscosity directly. Fig. 9 shows the related friction coefficient depending on viscosity for the viscosity range covered by engine oils SAE 5W to SAE 50 at 99 °C. It may be recognized that reducing the viscosity by one viscosity grade will result in a friction coefficient reduction between 13.2 and 18.5 %. It follows from Fig. 2 that this friction reduction can lead to a theoretically possible fuel consumption reduction between 0.6 and 5.5 % if 2/3 fluid film and 1/3 mixed film lubrication are taken into account.

Fig. 10 shows the same relationship for engine oils at -18 °C.

Obviously friction reductions of up to 30 % are realistic. According to Fig. 2 theoretically possible fuel consumption reductions between 1.0 and 7.5 % can be expected for the assumptions mentioned above.

4.2.2 Gear oils

From Fig. 9 (high temperature viscosity) and Fig. 10 (low temperature viscosity) also the effect of viscosity reduction of gear oils on friction can be estimated. While viscosity reduction by one SAE viscosity grade at high temperatures will result in friction reductions up to 24 %, this improvement will increase to about 45 % at low temperatures. Provided that about 50 % of friction between meshing gears is fluid friction - and higher values are not to be expected - according to Fig. 3 possible maximum reductions in fuel consumption between 0.2-0.4 % (favourable conditions) and 1.5-2.5 % (unfavourable conditions) are realistic.

4.3 Influence of friction reducing additives (friction modifiers)

Any similar numerical calculation of friction reduction by additives in the moment seems not to be possible. Therefore all considerations are limited to estimations of the order of magnitude of all effects. As it can be seen from Fig. 11 (22) in the mixed film lubrication regime a 50 % reduction of friction is possible if instead of pure mineral oils lubricants containing surface active additives are used. But it should be mentioned that exclusive the very right hand part of this diagram for these considerations is important. The extent of any friction reduction by an engine oil with special friction modifiers will be much lower, and a value of about 30 % seems to be realistic. With the assumption that only about 1/3 of the total friction in engines is mixed film friction according to Fig. 2 theoretical possible fuel consumption reductions between 0.7 and 4.0 % can be expected.

Regarding the larger parts of mixed film lubrication regimes in gears maximum fuel consumption reductions between 1.0 and 6.0 % could be taken into account.

5 POSSIBLE CONSUMPTION REDUCTIONS UNDER PRACTICAL CONDITIONS

In the following sections the results of an analysis of the maximum theoretical possible fuel consumption reduction by suited engine and gear oils as a target shall be discussed (17, 18).

5.1 Engine lubrication

5.1.1 General relationships

The following considerations are valid taking into account the following assumptions:

- Engine efficiency as shown in Table 2
- As an average 2/3 fluid film frictions and 1/3 mixed film or boundary friction are prevailing
- Friction reduction of 50 % by using other oils (for fluid film conditions this is more than discussed above).

Table 2: Efficiencies in OTTO and DIESEL engines

	EFFICIENCY	
	Total	Mechanical
OTTO-Engine		
Full Load	22 - 30	82 - 92
Partial Load	8 - 15	30 - 60
DIESEL-Engine		
Full Load	32 - 40	78 - 86
Partial Load	25 - 32	65 - 70

The specific full consumption values to overcome the mechanical losses are listed in Table 3 for different engine efficiencies. These figures are identical with the maximum possible reductions if the mechanical efficiencies could be raised to 100 %. Because this is impossible the relative small possibly existing improvements are obvious, especially under full load.

The highest fuel consumption reductions are to be expected under the following conditions (17):

- High mechanical losses, that means shortly after cold starts
- High speeds
- High loads.

Because it is unlikely that these conditions are prevailing simultaneously a further limitation of the practical fuel economy improvement is given.

Table 3: Specific fuel consumption by other mechanical efficiencies than 100 %

	η_{tot} in %	η_m in %	$\frac{\eta_{tot}}{\eta_m}$ in %	b_{e_m} in %
OTTO-Engine at Full load	22	92	23.9	1.9
	30	92	32.6	2.6
	22	82	26.8	4.8
	30	82	36.6	6.6
OTTO-Engine at Partial load	15	60	25.0	10.0
	8	40	20.0	12.0
	8	30	26.7	18.7
	15	40	37.5	22.5
DIESEL-Engine at Full load	32	86	37.2	5.2
	40	86	46.5	6.5
	32	78	41.0	9.0
	40	78	51.3	11.3
DIESEL-Engine at Partial load	25	70	35.7	10.7
	25	65	38.7	13.5
	32	70	45.7	13.7
	32	65	49.2	17.2

5.1.2 Representative driving programs

Now it will be attempted to link the calculated and estimated consumption reductions with the driving conditions of a realistic car population. The following considerations are based on three different driving programs representative for the Federal Republic of Germany (23). These programs are characterized in Fig. 12.

5.1.3 Percentage consumption reductions related to different driving conditions and programs

The estimation of the percentage consumption reductions related to the different driving conditions characterizing the driving programs of Fig. 12 are based on the following presumptions:

- Short distance traffic (Cold start and city traffic):
  Partial load with warm up phase of the engine, characterized by higher reductions at the beginning of this phase

- Intermediate distance traffic:
  1/4 full load and 3/4 partial load, characterized by intermediate to low reductions

- Long distance traffic:
  Mostly full load, characterized by low reductions

Taking into account these distribution of partial and full load conditions the maximum full consumption reductions according to Table 4 can be expected. These figures can be transferred to the different driving programs, linking the data of Fig. 12 and Table 4 resulting in Table 5 showing the maximum possible fuel consumption reductions for the different driving programs. Obviously for OTTO engines higher reductions are possible compared to DIESEL engines. As expected those programs characterized by more short distance traffic will result in higher reduction values.

Table 4: Maximum possible fuel consumption reduction at different driving conditions by other engine oils

Engine type \ Driving conditions	Short distance		Intermediate distance	Long distance
	Cold start	City traffic		
OTTO	7.5 %	6.0 %	2.2 %	1.8 %
DIESEL	5.8 %	4.8 %	2.5 %	3.0 %

Table 5: Possible fuel consumption reductions at different driving programs by other engine oils

Engine type \ Driving programs	A	B	C
OTTO	5.8 %	4.9 %	2.7 %
DIESEL	4.8 %	3.6 %	3.2 %

5.2 Gear lubrication

5.2.1 General relationships

The following considerations are based on the following assumptions:

- Gear efficiencies as discussed in chapter 3.2.2

- As an average not more than 50 % fluid film friction between meshing gears

- Friction reduction of 50 % by other oils

5.2.2 Representative driving programs

The same representative driving programs are taken into account as in the case of engine lubrication (Fig. 12).

5.2.3 Percentage consumption reductions related to different driving conditions and programs

Taking into account again the distribution of partial and full load conditions described above (chapter 5.1.3) the maximum full consumption reductions according to Table 6 can be expected. These figures can be transferred to the different driving programs, linking the data of Fig. 12 and Table 6 resulting in Table 7 showing the maximum possible fuel consumption reductions for the different driving programs. These calculated values are valid for the ideal case if high viscosity monograde oils without friction modifier are substituted by low viscosity multigrade oils containing friction modifier.

Table 6: Maximum possible fuel consumption reduction at different driving conditions with modified gear oils

	Short distance		Intermediate distance	Long distance
	Cold start	City traffic		
Range of maximum possible fuel consumption reduction	% 1.4-6.3	% 1.3-4.9	% 1.2-4.1	% 0.9-2.9

Table 7: Possible fuel consumption reductions at different driving programs with modified gear oils

	Driving program A	B	C
Range of maximum possible reductions	1.3-5.1 %	1.2-4.5 %	1.0-3.6 %

5.3 Total fuel consumption reduction

The total fuel consumption reductions are given by the sum of the single reductions by modified engine or gear oils. These data are summarized in Table 8.

Table 8: Summarizing the calculated theoretically possible maximum reductions

	Range of maximum reductions: Theoretically possible	Considering different driving programs
Engine oils	1.8 % - 7.5 %	2.7 % - 5.8 %
Gear oils	0.9 % - 6.3 %	1.0 % - 5.1 %
Engine and Gear oils	2.7 % - 13.8 %	3.7 % - 10.9 %

6 CONCLUSION AND SUMMARY

The results of these considerations may be summarized and evaluated as follows:

a. First of all it has to be considered that the fuel consumption of a car depends on a set of parameters that related to lubricants. Their influence mostly is much more pronounced than the lubricant influence.

b. By lubricant related measures only the mechanical losses can be decreased. Therefore the fuel economy improvement which possibly might be realized are rather limited.

c. It has to be taken into account that only about 1/3 of the total friction losses occur in the mixed film or boundary regime, whereas 2/3 are fluid film friction losses. This ratio has to be kept in mind, if the relative influences of friction modifier or lower viscosities are estimated and evaluated. Changing viscosities will vary this ratio.

d. Evaluating the viscosity influence on fuel consumption the so-called effective viscosity has to be taken into account. This is most important for non-newtonian oils.

e. Reducing the engine oil viscosity by one SAE viscosity grade will result in fuel consumption reductions of 0.6 to 5.5 % at high temperatures and 1.0 to 6.5 % at low temperatures. The corresponding data for gear oils are 0.2 to 1.5 % (high temperatures) and 0.4 to 2.5 % (low temperatures).

f. By using friction modifiers in engine oils, fuel consumption reductions between 0.7 and 4.0 % and in gear oils reductions between 1.0 and 6.0 % are realistic.

g. On the basis of a 50 % reduction of friction maximum fuel consumption reductions between 2.7 and 5.8 % by modified engine oils and between 1.0 and 5.1 % by modified gear oils are possible dependent upon the driving programs. The total reduction is between 3.7 and 10.9 %.

h. Considering the relative distribution of the different driving programs, the car population as well as the driving behaviour in the Federal Republic of Germany fuel consumption reductions up to 4.6 % by other engine oils and up to 2.8 % by other gear oils are realistic. The maximum total reductions on this basis are 7.4 %. This would result in an annual reduction of mineral oil consumption of 1.3 % - a rather small figure.

Other results would be obtained taking into account the conditions in the United States of America.

7 LITERATURE

(1) BARTZ, W.J. Gedanken zur Kraftstoffeinsparung durch reibungssenkende Motoren- und Getriebeöle. Arbeitskreis Schrifttumsauswertung Schmierungstechnik, Bericht Nr. 01/02-81

(2) BARTZ, W.J. Kraftstoffeinsparung durch Reibungsminderung bei Motoren- und Getriebeölen. MTZ Motortechnische Zeitschrift 41 (1980) 1, 7-12

(3) BARTZ, W.J. Mögliche Energieeinsparung durch tribologische Maßnahmen. Erdöl und Kohle 33 (1980) 2, 78-87

(4) BARTZ, W.J. Mögliche Energieeinsparung durch tribologische Maßnahmen. Arbeitskreis Schrifttumsauswertung Schmierungstechnik, Bericht Nr. 13/14-79

(5) BARTZ, W.J. Improving fuel economy by friction reducing engine and gear oils. Proc. AGELFI European Automotive Symposium 313-342, 1980

(6) TURNBULL, K.H., G.S. LEITHEAD Fuel economy characteristics of motor oils as measured in a consumer road test. SAE Paper 79 09 46

(7) PAPAY, A.G., E.B. RIFKIN, R.L. SHUBKIN, P.F. JACKISCH, R.B. DAWSON Advanced fuel economy engine oils. SAE Paper 790947

(8) LONSTRUP, T.F., H.E. BACHMANN, C.R. SMITH Testing the fuel economy characteristics of engine oils. SAE Paper 790949

(9) WILLETTE, G.L. Evaluation of a fuel efficient long drain ester based lubricant. SAE Paper 790950

(10) JONES, E.R., L.J. PAINTER Some statistical aspeccs of testing engine oils for fuel economy. SAE Paper 790951

(11) O'CONNOR, B.M., R. GRAHAM, I. GLOVER European experience with fuel economy gear oils. SAE Paper 790746

(12) MC'GEEHAM, J.A. A literature review of the effect of piston and rig friction and lubricating oil viscosity on fuel economy. SAE Paper 780673

(13) BADIOLI, F.L., A.A. CASSIANI-INGONI, G. PUSATERI Friction power loss of mineral and synthetic lubricants in a running engine. SAE Paper 780376

(14) MC'DONELL, T.F., S.A. TEMPE The effects of engine oil additions on vehicle fuel economy, emulsions, emulsion control components and engine wear. SAE Paper 780962

(15) HAVILAND, M.L., M.C. GOODWIN Fuel economy improvements with friction-modified engine oils in environmental protection agency and road tests. SAE Paper 790945

(16) PINKUS, O., D.F. WILCOCK The role of tribology in energy conservation. Lubrication Engineering 34 (1978) 11, 599-610

(17) KASPER, G. Möglichkeiten zur Kraftstoffeinsparung bei Verbrennungsmotoren durch Verwendung anderer Schmierstoffe. Studienarbeit, Universität Stuttgart, Institut für Verbrennungsmotoren und Kraftfahrtwesen, Fachgebiet Tribologie, 1979

(18) WANNER, G. Möglichkeiten zur Kraftstoffeinsparung bei Kraftfahrtzeugen durch Verwendung anderer Schmierstoffe in den Stufen- und Achsgetrieben. Studienarbeit, Universität Stuttgart, Institut für Verbrennungsmotoren und Kraftfahrtwesen, Fachgebiet Tribologie, 1979

(19) ROWE, C.N. Reduction of Engine Friction losses via Lubricant Chemistry and Physical Properties. Gordon Research Conference "Friction, Lubrication and Wear", New London, N.H.

(20) GEORGI, C.W. The effects of motor oils and additives on engine fuel consumption. SAE Transactions 62 (1954), 385-391

(21) PASSUT, C.A., R.E. KOLLMANN Laboratory Techniques for Evaluation of Engine Oil Effects on Fuel Economy. SAE Paper 780601

(22) RABINOWICZ, E. Surface Energy Approach to Friction and Wear. Product Engineering (1965), 15. März, 95-99

(23) DABELSTEIN, W. Der Energiebedarf des Straßenverkehrs unter Berücksichtigung steigender Anteile am Diesel-Personenwagen. Motortechn. Z. 38 (1977) 12, 113-116

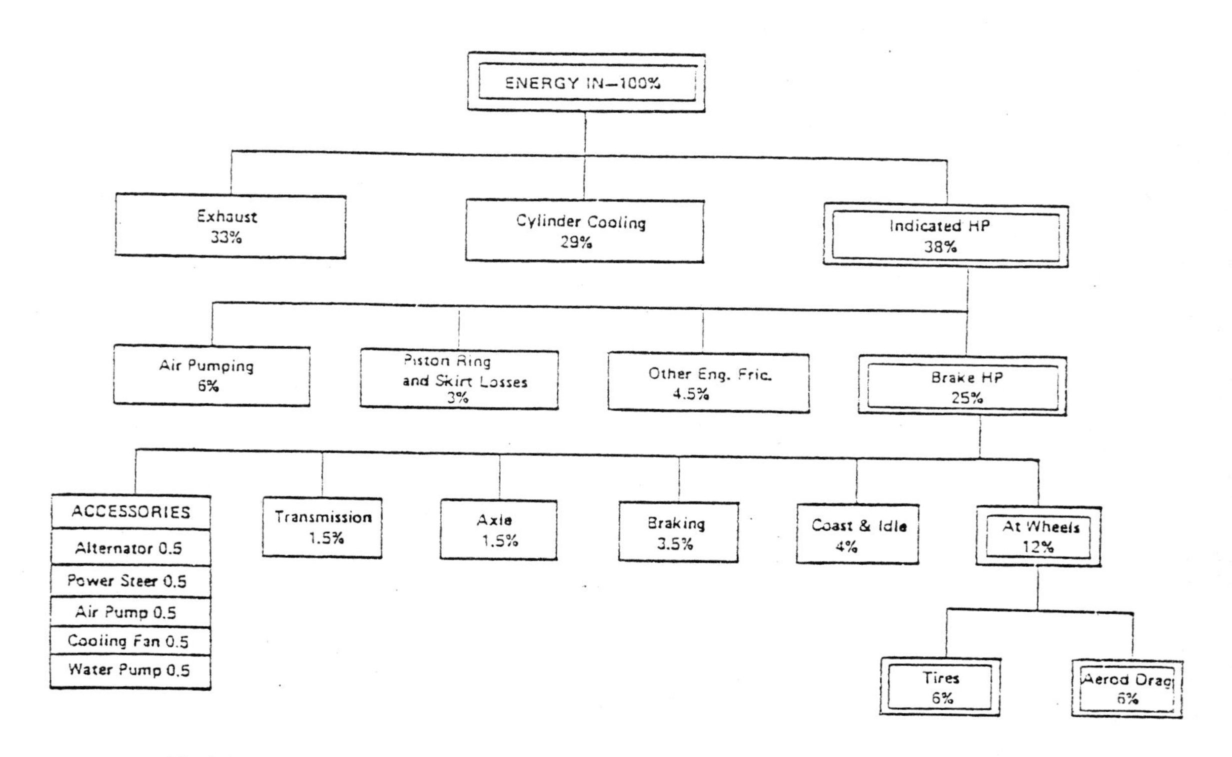

Fig 1 Energy distribution in a passenger car during a EPA-city/highway-cycle

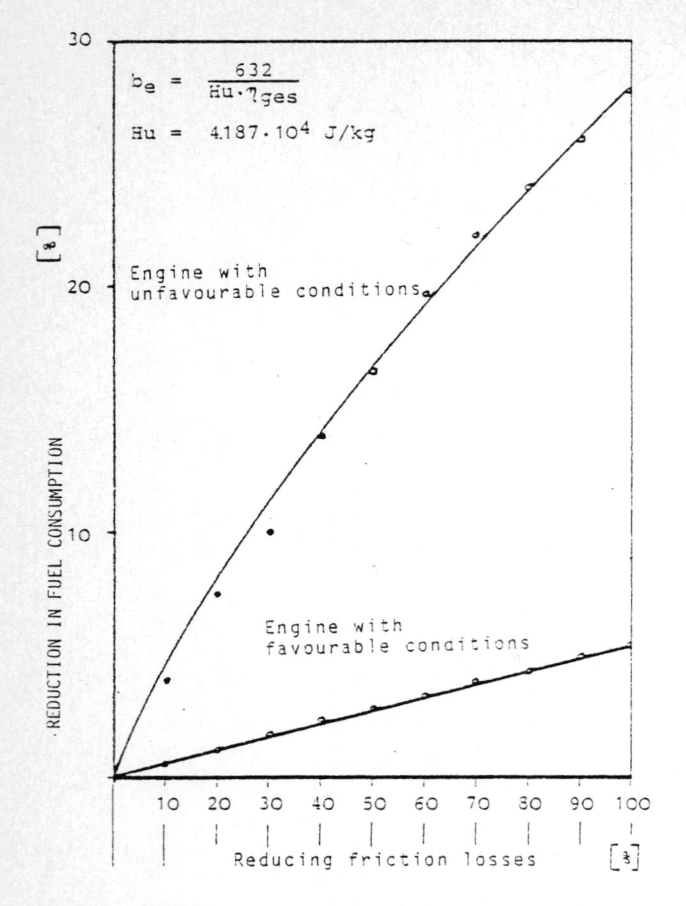

Fig 2 Reduction in friction losses depending on friction reduction

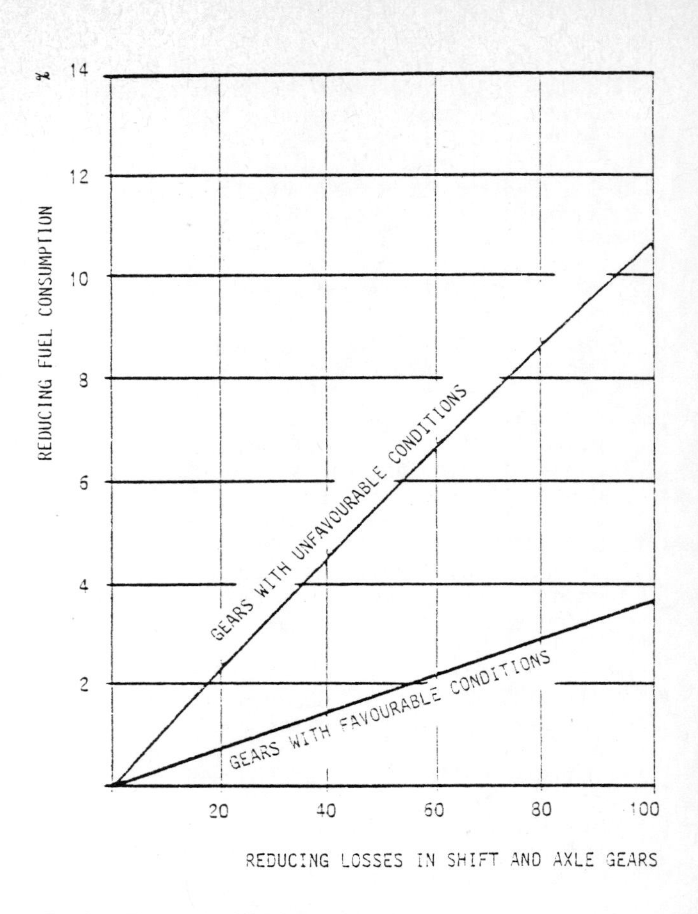

Fig 3 Theoretical fuel consumption reduction by reducing losses in shift and axle gears (according to Wanner (18))

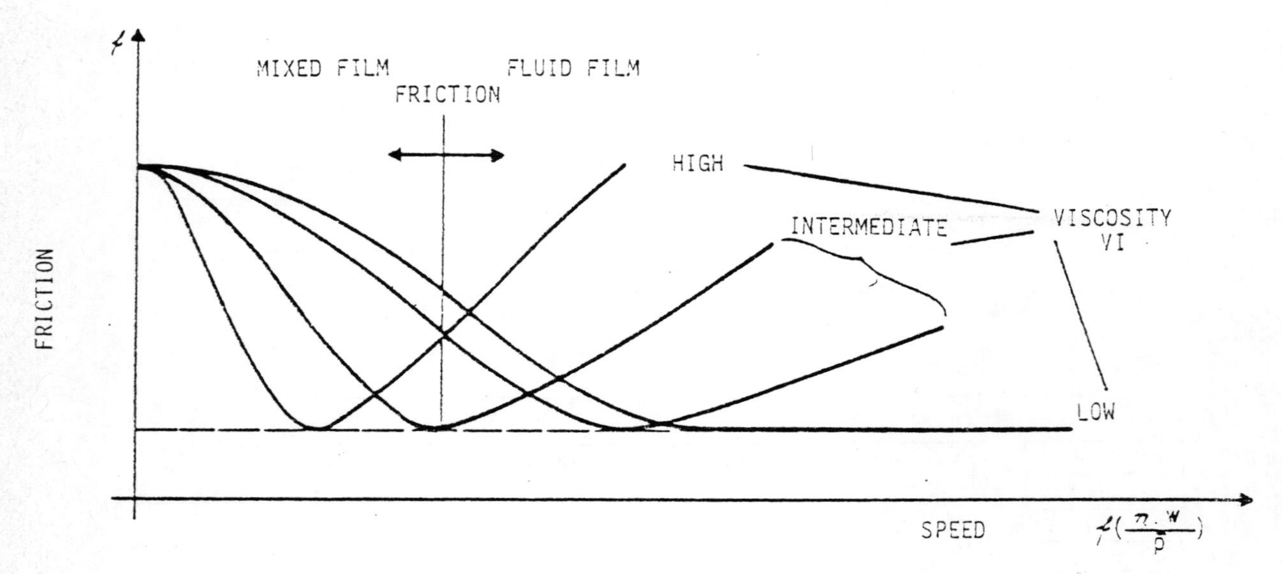

Fig 4 Friction depending on speed for different viscosities (schematically)

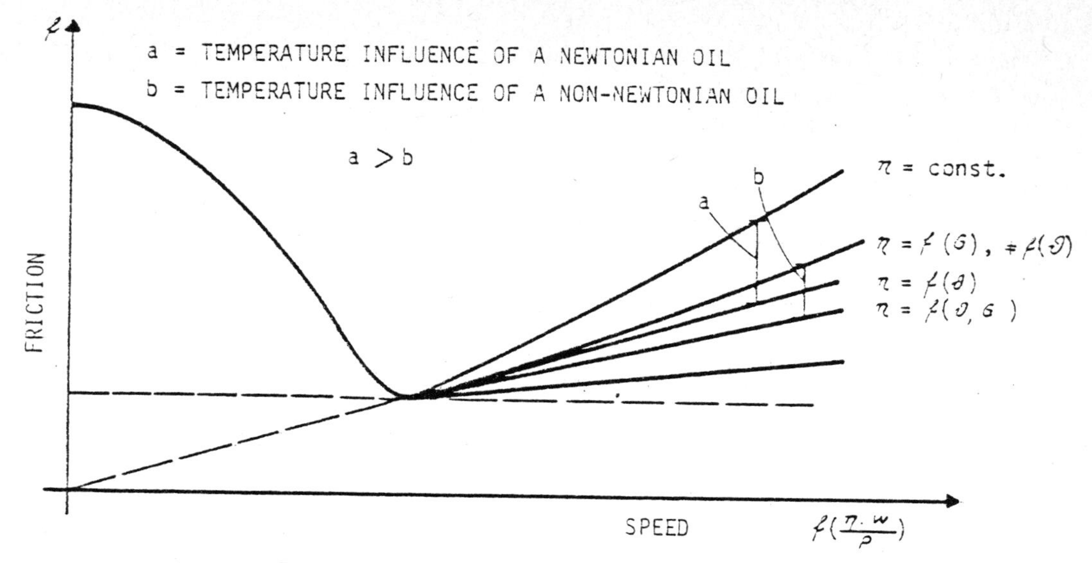

Fig 5 Friction depending on speed for oils with different viscosity behaviour (schematically)

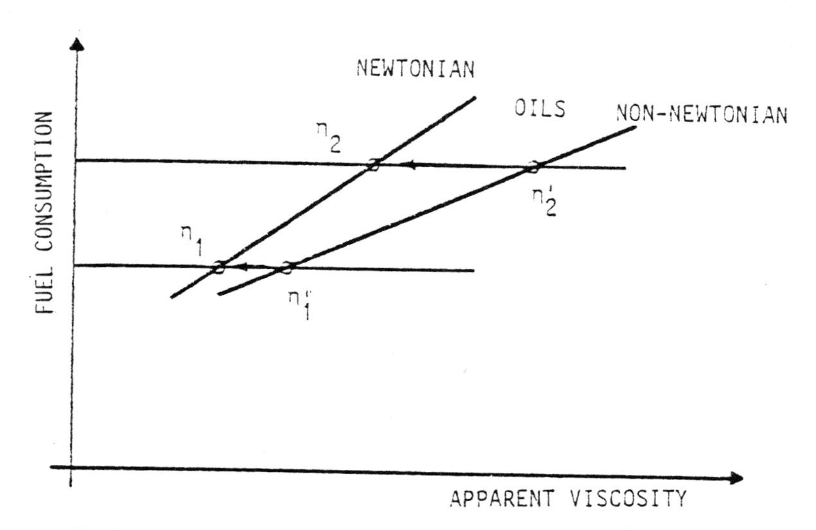

Fig 6 Fuel consumption depending on viscosity for Newtonian and non-Newtonian oil (schematically)

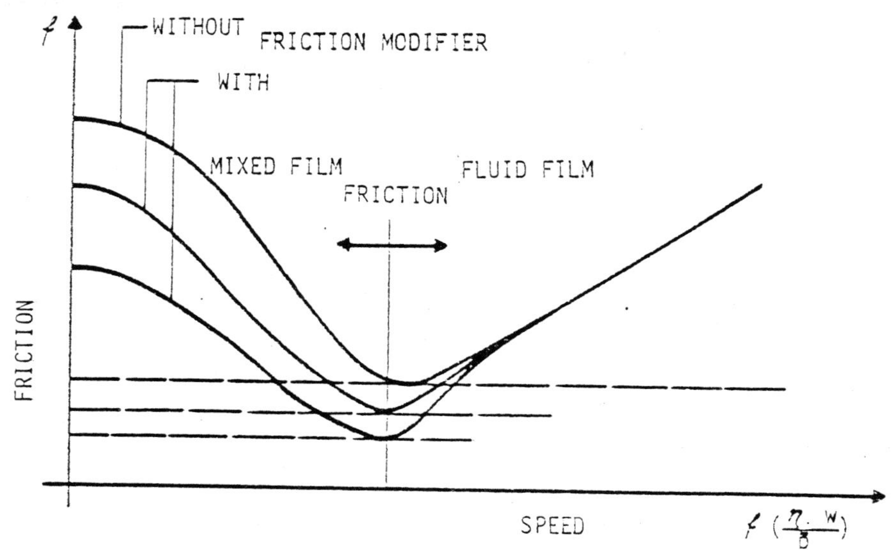

Fig 7 Friction depending on speed for oils having different boundary lubrication behaviour (schematically)

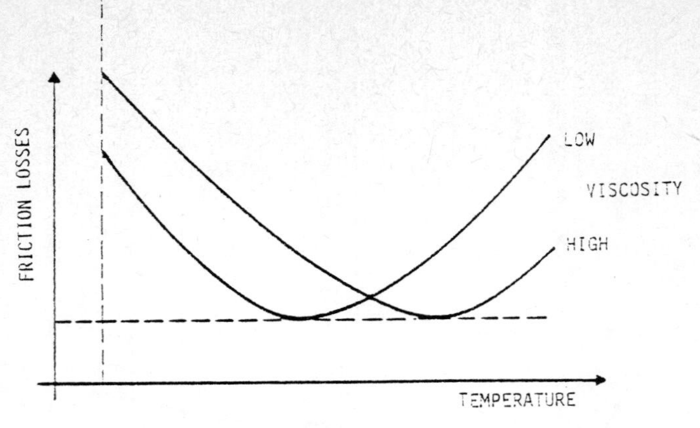

Fig 8 Friction losses depending on temperature for oils having different viscosities (schematically)

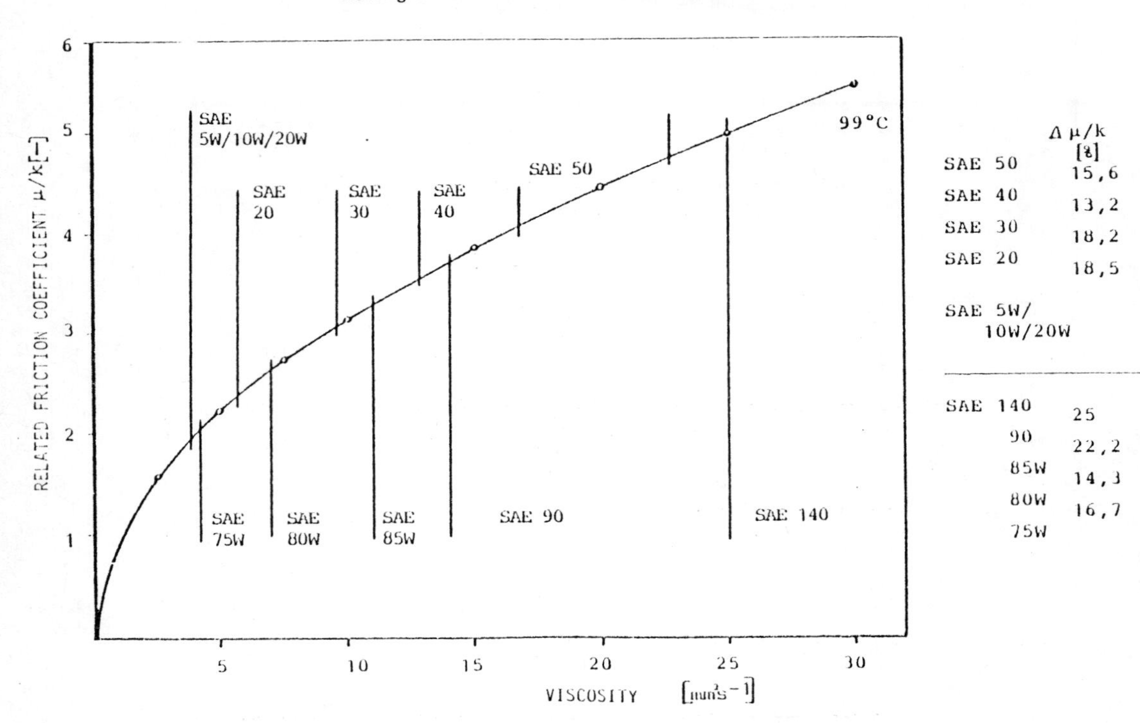

	Δ μ/k [%]
SAE 50	15,6
SAE 40	13,2
SAE 30	18,2
SAE 20	18,5
SAE 5W/10W/20W	
SAE 140	25
90	22,2
85W	14,3
80W	16,7
75W	

Fig 9 Related friction coefficient depending on viscosity at 99°C

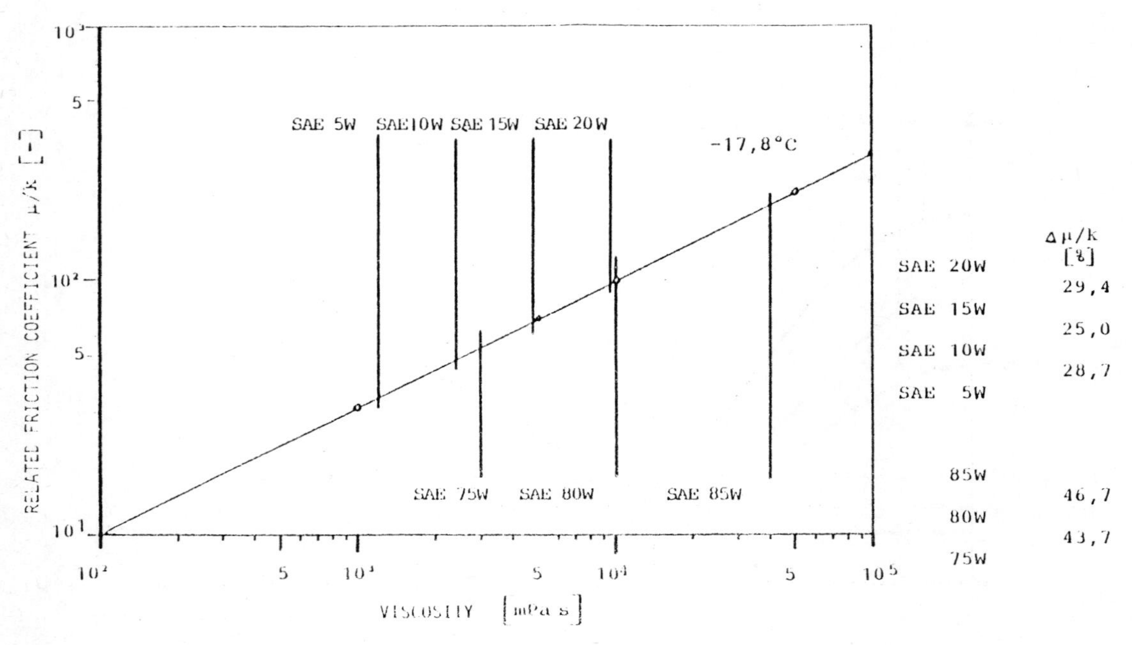

	Δ μ/k [%]
SAE 20W	29,4
SAE 15W	25,0
SAE 10W	28,7
SAE 5W	
85W	46,7
80W	43,7
75W	

Fig 10 Related friction coefficient depending on viscosity at −18°C

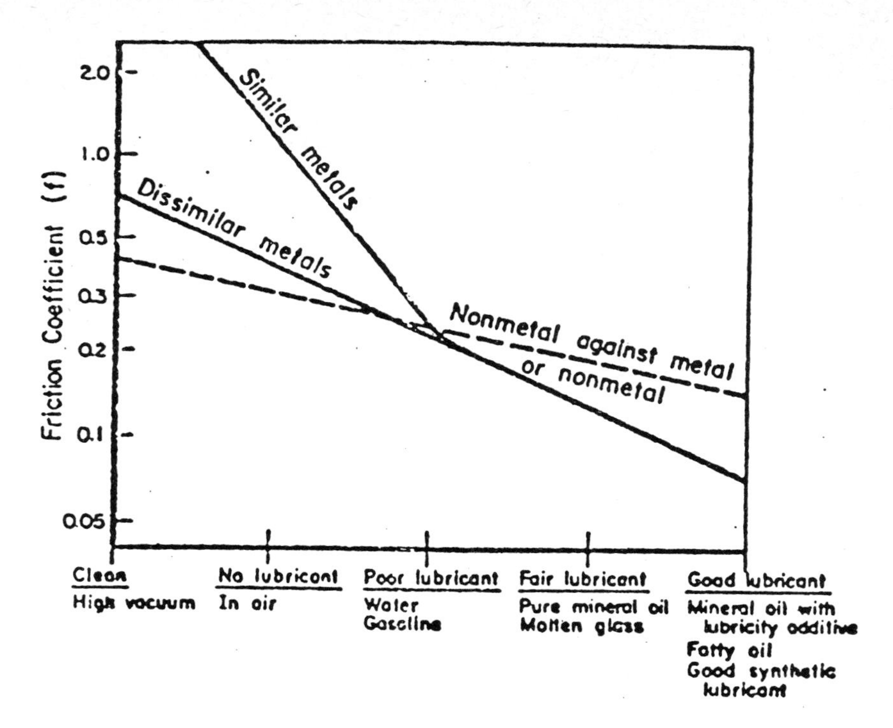

Fig 11 Friction coefficient depending on lubrication regime

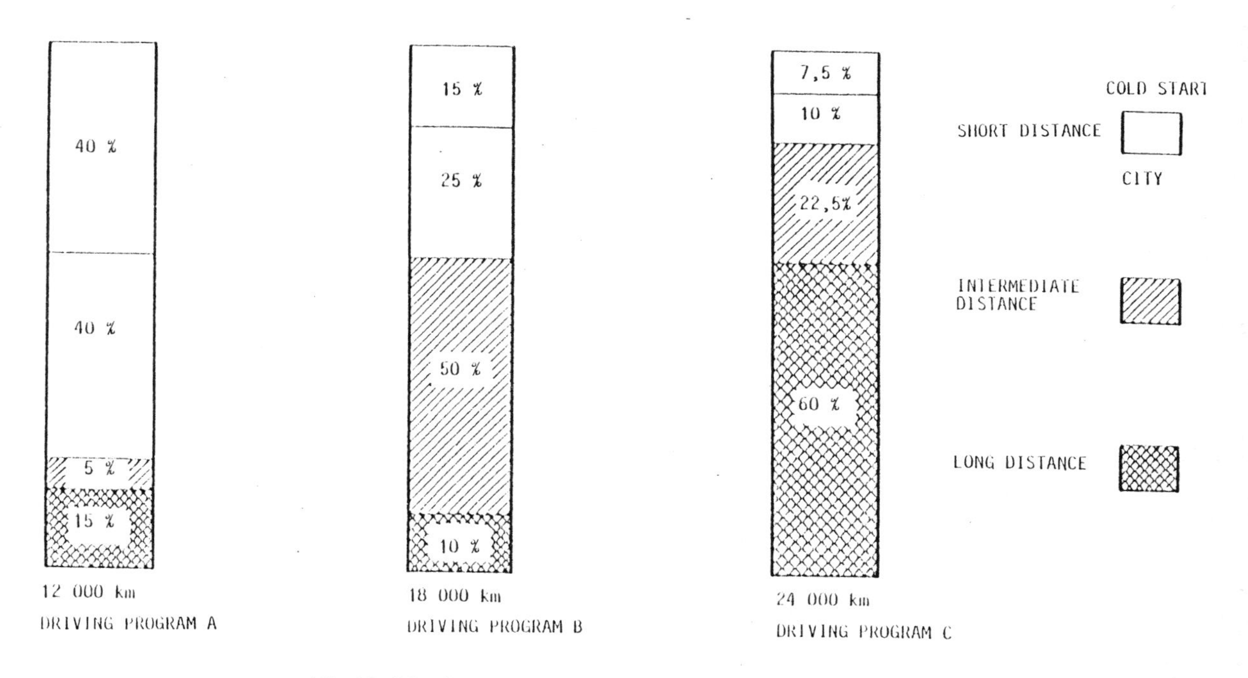

Fig 12 Distribution of three representative driving programs on short distance, intermediate distance, and long distance traffic

C61/85

Measurement and characteristics of instantaneous piston ring frictional force

K SHIN and Y TATEISHI
Nippon Piston Ring Company Limited, Saitama-Pref., Japan
S FURUHAMA
Musashi Institute of Technology, Tokyo, Japan

SYNOPSIS An actual engine and the rig tester were used to find general characteristics of the piston assembly and piston ring frictions, and measurements of the instantaneous frictional force and the instantaneous oil film thickness were carried out. According to test results of the rig tester, the measured oil film thickness and the calculated value as a complete fluid lubrication agreed fairly well, while the measured frictional force was greater by twofold than the calculated value. It was assumed that the frictional loss process for the running-in in a new ring in a new cylinder should have the optimum value for the cylinder bore surface roughness.

1. INTRODUCTION

Improved fuel economy of internal combustion engines is a major objective of research efforts these days. Reducing the frictional loss between piston rings and cylinders is an especially useful means to attain it.
Generally speaking, measures for lessening frictional losses in piston operation bring on the reduction of lubricant oil film thickness. This means a greater metal-to-metal contact and engine wear, resulting in a loss of endurance. To prevent this problem, it is necessary to develop a countermeasure that will reduce friction while avoiding the risk of non-fluid lubrication as much as possible. For this purpose, the conventional friction measuring method, that is, the motoring method and indicated mean effective pressure (IMEP) method,(1) are inadequate by themselves because they show only the mean of friction in one cycle of operation. For instance, lubricants of lower viscosity and piston rings of smaller axial breadth certainly reduce friction but suffer heavier wear. It is impossible to consider the extent of the latter from the mean value. In other words, it is a must to measure changes in the frictional force and lubricant oil film thickness during one cycle of operation.
Hamilton et al(2) of Reading University measured the lubricant oil film thickness on piston rings by attaching a small electrode to the cylinder, and pointed out that the measured thickness is smaller than the theoretical one. They supposed that this gap between the measured and theoretical oil film thicknesses was caused by an inadequate supply of lubricants.
This paper aims to describe general characteristics of piston frictional losses by measuring the instantaneous frictional forces and instantaneous oil film thicknesses of the piston assembly.

2. FRICTIONAL FORCE AND OIL FILM THICKNESS MEASURING DEVICES AND METHOD

A rig tester and actual operating engine were tested.
The Cylinder liner floating method(3),(4) was used in measuring frictional force actual operating engines operation. With the rig tester the piston rings and piston were held stationary and the cylinder liner was set in reciprocating motion.(5) A diesel engine crank mechanism was used in the tester, and a moving cylinder of ∅80 mm bore was installed in the place of a piston. A cylinder guide was attached to the top of the cylinder block. The piston rings and piston were suspended from the top center of the cylinder guide through a load cell.
The moving cylinder reciprocated at a stroke of 102 mm in order to measure the frictional force of piston rings.
An electric capacitance pickup whose surface area was 0.39 mm^2 was attached to the piston ring sliding surface for simultaneous measurement of oil film thicknesses and frictional force.
A heater was located on the outside of the cylinder guide to adjust lubricant oil temperature on the piston ring sliding surface.
The lubricant used was a single grade oil of SAE - 30, and it was fully supplied around the piston ring.
An electric capacitance method was also used to measure the oil film thickness of the operating diesel engine in the same way as the rig tester. As the oil film thickness on the operating diesel engine was supposed to be considerably small, the surface of the aluminium electrode was subjected to aluminium oxide treatment for insulation.
A link mechanism(6) was used to lead out the lead wires from the piston rings to the outside of the engine.

3. TYPICAL EXAMPLES OF RESULTS OF MEASUREMENT

Fig. 1 (a) shows the results of oil film thickness and frictional force measurement with the rig tester at 1,000 rpm and oil temperature of 80°C. Diagram B on the top indicates oil film thickness and A shows the presence of a calibration groove which was provided to aid in finding the absolute value of oil film thickness.

Oil film thicknesses were found from calibration signals.
The top dead center of the moving cylinder lies between the peaks. The oil film becomes thickest at the maximum speed point near the center of stroke, and is thinnest immediately after the top and bottom dead centers. A minimum value is indicated for the latter.
These characteristics coincide with fluid lubrication characteristics under a constant piston ring-cylinder contact load.
D in the bottom indicates frictional force, which is almost zero at the top and bottom dead centers and becomes greatest near the center of stroke.
A maximum frictional force F0 generated at near the center of stroke was 12N. The 0-line shown is the center line of 2F0.
These results are thought to bear out the supposition that piston rings are in a state of almost complete fluid lubrication, when the oil is supplied adequately without gas pressure.
Fig. 1 (b) is a typical example of results of frictional force measurement of piston assembly of a small diesel engine. F is a friction diagram, and F0 indicates a maximum frictional force without gas pressure in the intake and exhaust strokes. Fm is a maximum frictional force immediately after the combustion TDC under gas pressure.
Pg is combustion pressure and tc is cylinder bore temperature at the midpoint of top ring travel.

4. TEST RESULTS

4.1 Rig test

Fig. 2 is a comparison of oil film thickness and frictional force values which were measured while changing the piston ring axial breadth on the rig tester, and theoretical values calculated according to the hydrodynamic lubrication theory(7) advocated by Furuhama. The piston rings were tested at axial breadth (B) of 1.2 mm and 2.0 mm, piston ring tension W_T of about 10N, operating speed of 1,000 rpm and oil temperature 80°C.
The oil film thickness is reduced by narrowing the axial breadth of piston rings, and its absolute values coincide closely with theoretical values.
The frictional force did not change because tension, that is, contact load (B.Pe = constant, Pe: elastic contact pressure) was kept constant regardless of the change in axial breadth. This coincides with the result of fluid lubrication theories. But the absolute value of frictional force is almost twice as much as the theoretical value.

4.2 Relation between F0 and friction mean effective pressure Pf

The relation between F0 and Pf indicating frictional loss in the frictional force curve concerning the actual piston assembly of actual operating engines, is shown in Fig. 3. It is evident that Pf is interrelated to F0 regardless of load conditions. The frictional force gains markedly under high load near TDC where gas pressure is high, but its effect is minor, compared with overall frictional losses.

These results show that, in order to confirm the frictional loss in piston operation, firing of actual piston assemblies is not a must but motoring or rig tests with a reciprocating cylinder are just as useful, provided that oil temperature and piston speed be adjusted to be same as firing.
In order to confirm the state of lubrication, it is necessary to know the value of Fm. To find the value of Fm, it is useful to make a firing test using actual piston assemblies for frictional force measurement.

4.3 Frictional force measurement with one piston ring using the rig tester and diesel engine in motoring

The frictional force per unit length (R) at the maximum piston speed point during intake stroke of a piston ring of diesel engine in the motoring, and its rig tester counterpart were arranged in order by friction coefficient f and T value (function of B/h, shape of piston ring sliding surface) for $W/6\mu U$ (μ: oil viscosity, U: Maximum sliding speed, W: B.Pe=load), and theoretical values and test values were compared. Results are shown in Fig. 4.
Toward the right of Fig. 4 where $W/6\mu U$ is large, R of the rig test is smaller than R detected in the diesel engine motoring test.
This is mainly attributable to the difference in estimating oil viscosity or temperature. Close matching in trend with line of theoretical values is evident, and absolute values are about twice as much.

4.4 Results of frictional force measurement with actual assembly

Fig. 5 shows results of frictional force measurement with the four piston rings one by one and assembled into an operating package.
The measured frictional force of individual pistion rings is about twice as much as the theoretical value as mentioned above.
The measured friction value of the piston ring package is about 1.7 times as much as the arithmetic sum of each ring's measurements.
An interference in oil distribution among the rings in the package, that is, oil starvation of each ring in the package, resulting in the reduction of oil film thickness and an increase of the frictional force, is thought to be the cause.

5. RESULTS OF OIL FILM THICKNESS MEASUREMENT WITH OPERATING DIESEL ENGINE

Oil starvation is thought to be a cause of producing piston ring frictional force measurements considerably greater than the theoretical value. In this connection, an actual diesel engine piston ring package of 140 mm bore and 152 mm stroke, consisting of three compression rings and one oil control ring, was used to measure the thickness of oil film on the top ring. Results are shown in fig. 6 (a), and considerably smaller than the theoretical value.
Measurement was also taken using only the top ring. Results are shown in Fig. 6 (b). Compared with the 4-ring package, oil film becomes considerably thicker when measured using only the

top ring, and measurements are almost all near the theoretical value regardless of local departure in the thicker or thinner direction from the theoretical value.

6. INFLUENCE OF CYLINDER SURFACE ROUGHNESS ON FRICTIONAL FORCE

A cast iron piston ring (B=2.0 mm, W_T=10N) was used on the rig tester, and the influence of cylinder surface roughness on piston ring frictional force was measured using new cylinders of 7 μm, 3 μm and 1 μm of surface roughness (R3tm) with new rings. The tester was operated for 7 hours at oil temperature of 100°C and piston speed 500 rpm.
Resultant changes in the frictional force diagram and fluctuations of mean effective frictional force Pf are shown in Fig. 7. The frictional force in 2 minutes is greater with surface roughness of 7 μm and 1 μm than 3 μm. Friction peaks at the top and bottom dead centers regardless of surface roughness, indicate occurrence of a metal-to-metal contact. At the end of 7 hours the frictional force peaks at the top and bottom dead centers are reduced for surface roughness of 3 μm and 1 μm. This indicates that a state of fluid lubrication is almost attained. When surface roughness is 7 μm, however, the metal-to-metal contact is still present at the top and bottom dead centers after 7 hours of operation.
As for Pf, the value becomes stable in two to three hours when surface roughness is 1 μm and 3 μm, but it tends to decrease even after 7 hours for surface roughness 7 μm, indicating continued running-in.
After 7 hours the piston ring sliding surface becomes a lubricating surface with symmetrical wear on each side. When cylinder surface roughness is 1 μm, however, an area of inadequate running-in is left in the center of the piston ring sliding surface. When cylinder surface roughness is finer, the lubricating surface is slower to form on piston rings. When tested at cylinder surface roughness of 3 μm and 7 μm, the lubricating surface was faster to develop on the ring than when cylinder surface roughness was 1 μm.
Regardless of the difference in surface roughness, the state of cylinder sliding surfaces after the test was such that asperities were flattened more or less.
Fig. 8 (a) shows the frictional force and oil film thickness after 7-hour operation at 500 rpm and oil temperature of 100°C. The frictional force at the maximum piston speed near the center of stroke is equal, regardless of surface roughness. Oil film thickness there is about 4 μm. For cylinder surface roughness of 7 μm, however, the frictional force near the top and bottom dead centers is heavy and oil film thickness is 2 μm or so there. This resulted in a metal-to-metal contact.
Therefore, measurement was taken by lowering oil temperature and raising piston speed (oil temperature: 80°C, piston speed: 1,000 rpm) so that a minimum oil film thickness of more than 4 μm or so might be attained. Results are shown in Fig. 8 (b).
The state of fluid lubrication occurs almost over the full range of stroke, and the frictional force was same for cylinder surface roughness of 7 μm and 3 μm.

Fig. 8 (c) shows results of measurement using an operating diesel engine. The frictional force peaks at the top and bottom dead centers in the initial phase of operation were the same as the rig test. It also took about 20 hours to complete running-in.

7. CONCLUSION

Frictional force and oil film thickness measurement using the rig tester and operating diesel engine disclosed the following.

(1) When measured using only one ring, the actual and theoretical oil film thicknesses coincide closely because of adequate oil supply. Frictional force changes show a similar tendency to that of theoretical values. But the absolute values are about twice as much.

(2) Results of frictional force measurement are same for the motoring diesel engine and the rig test, when they are arranged in order by W/6μU.

(3) Piston frictional loss Pf is interrelated to maximum frictional force F0 at the center of stroke without gas pressure in the frictional force curve. Therefore, when it is necessary to find only frictional loss, the rig test under same oil film temperature and piston speed conditions as the actual operating engine is useful in finding F0. But in order to evaluate wear resistance, it is necessary to find Fm by testing the actual operating engine.

(4) The overall frictional force of the four piston rings in combination is greater by 70% than the sum of frictional forces of individual rings. This is accounted for by oil starvation.

(5) As for the relation between cylinder surface roughness and frictional forces, the most favorable results are produced when the former is 3 μm. When cylinder surface roughness is finer or coarser, the initial frictional force is high and running-in takes time.

The authors express their sincere thanks to Mr. O. Mori and Laboratory stuff of Musashi Institute of Technology for their invaluable assistance and encouragement during preparation of this paper.

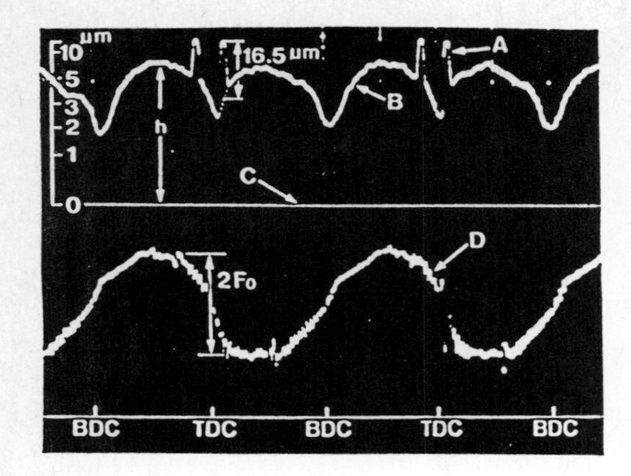

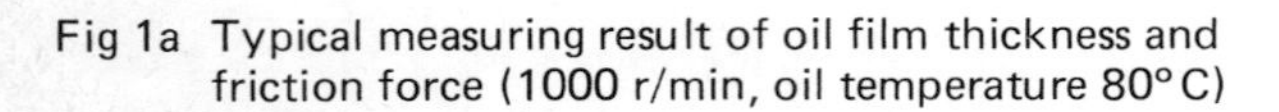

h : Oil film thickness

A : Peak for calibration

B : Change of oil film thickness

C : Cylinder surface (0-line)

D : Change of friction force

Fig 1a Typical measuring result of oil film thickness and friction force (1000 r/min, oil temperature 80° C)

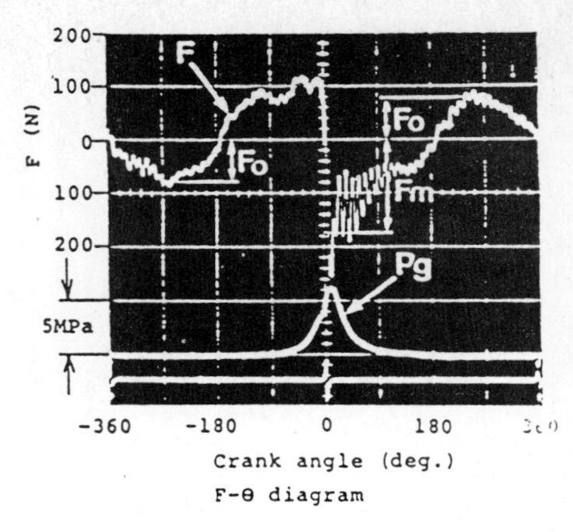

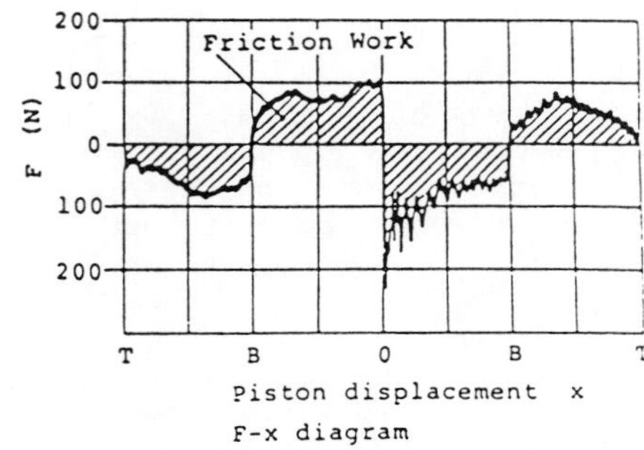

Fig 1b Typical piston friction force progress in diesel engine (1000 r/min, t_c = 90° C)

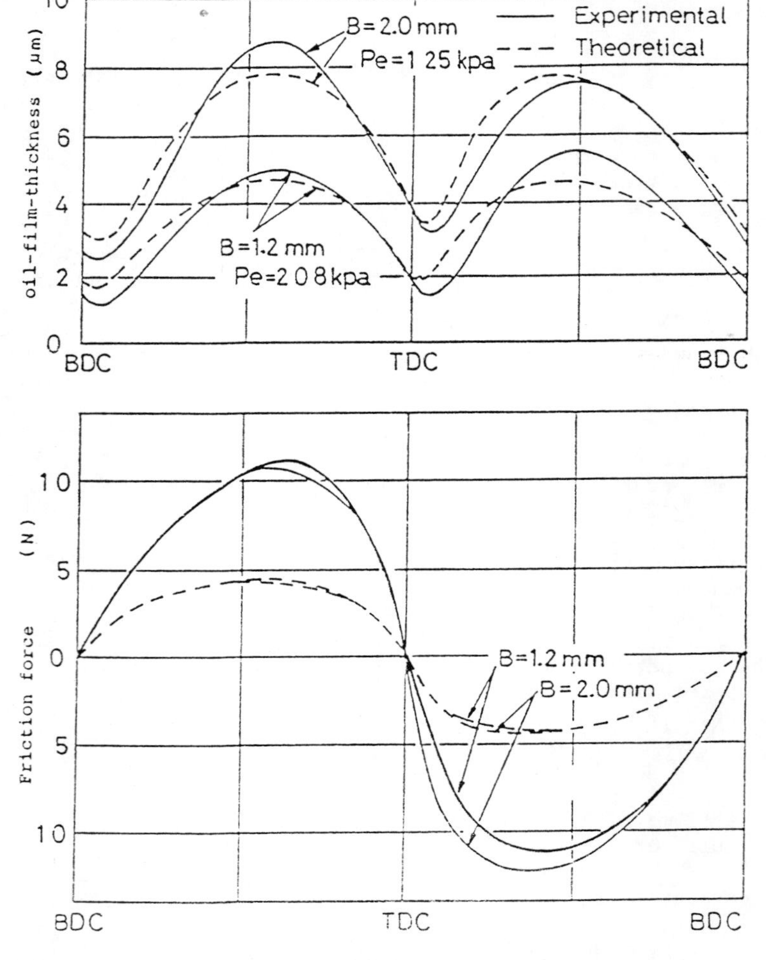

Fig 2 Comparison of measured value with the theoretical one when piston ring axial breadth was changed

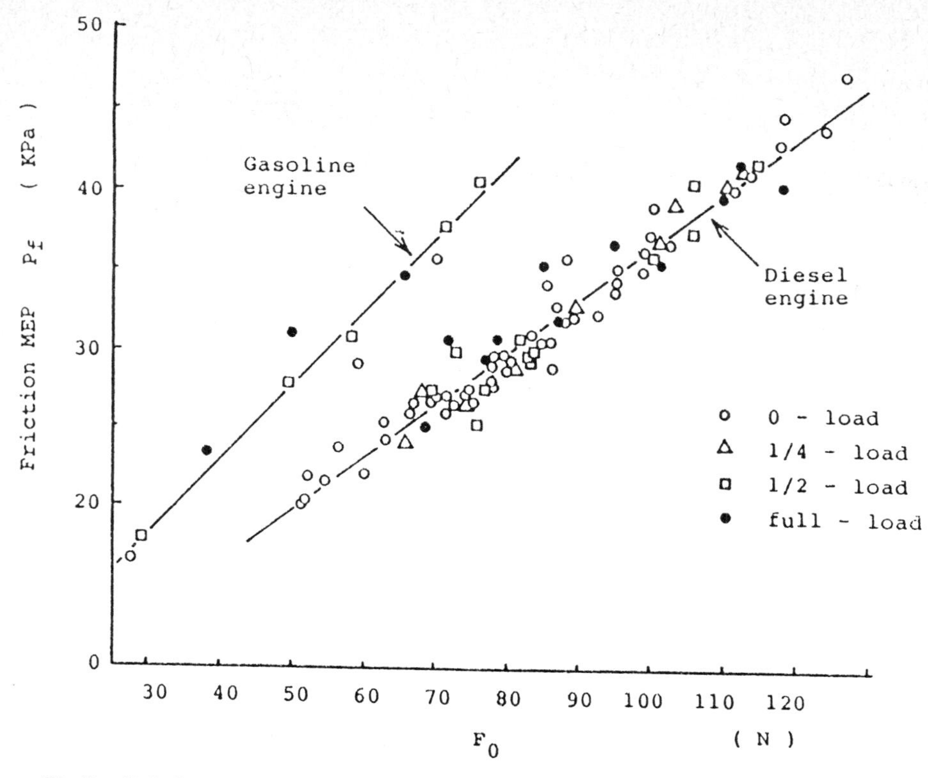

Fig 3 Relation between P_f and F_o at various operating conditions

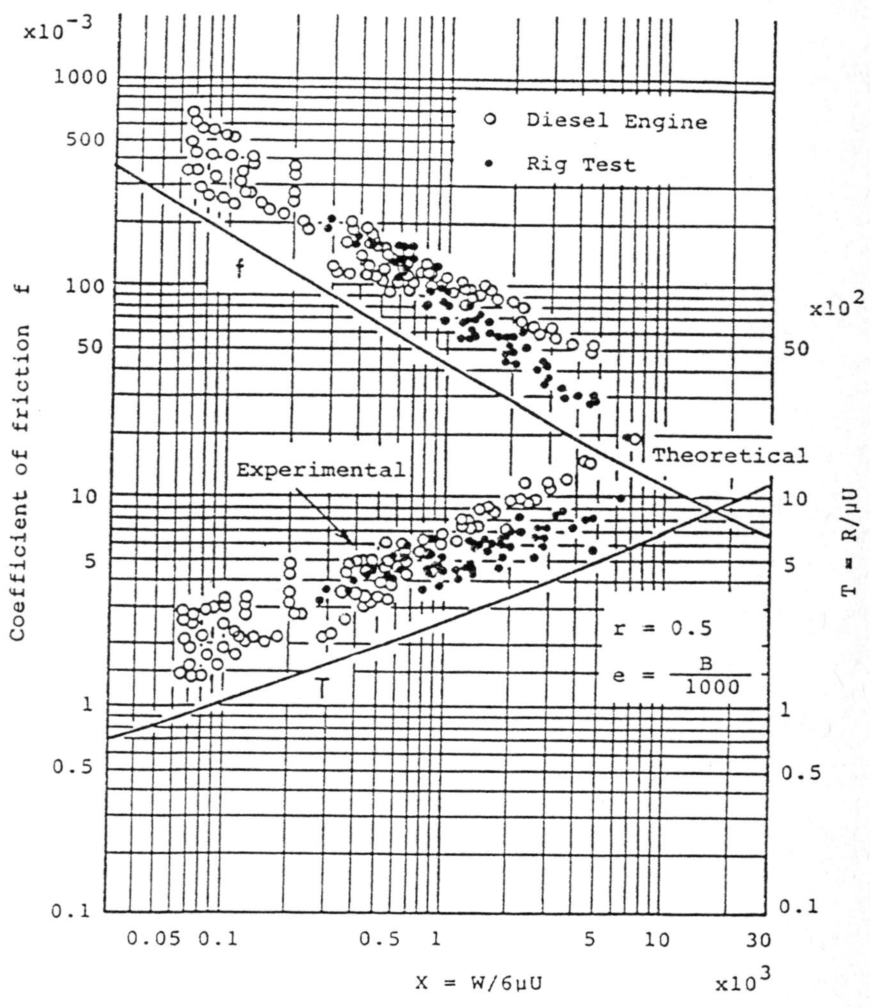

Fig 4 Relation between coefficient of friction f and lubricating condition (W/6 μU) when diesel engine in motoring condition and rig tester were used

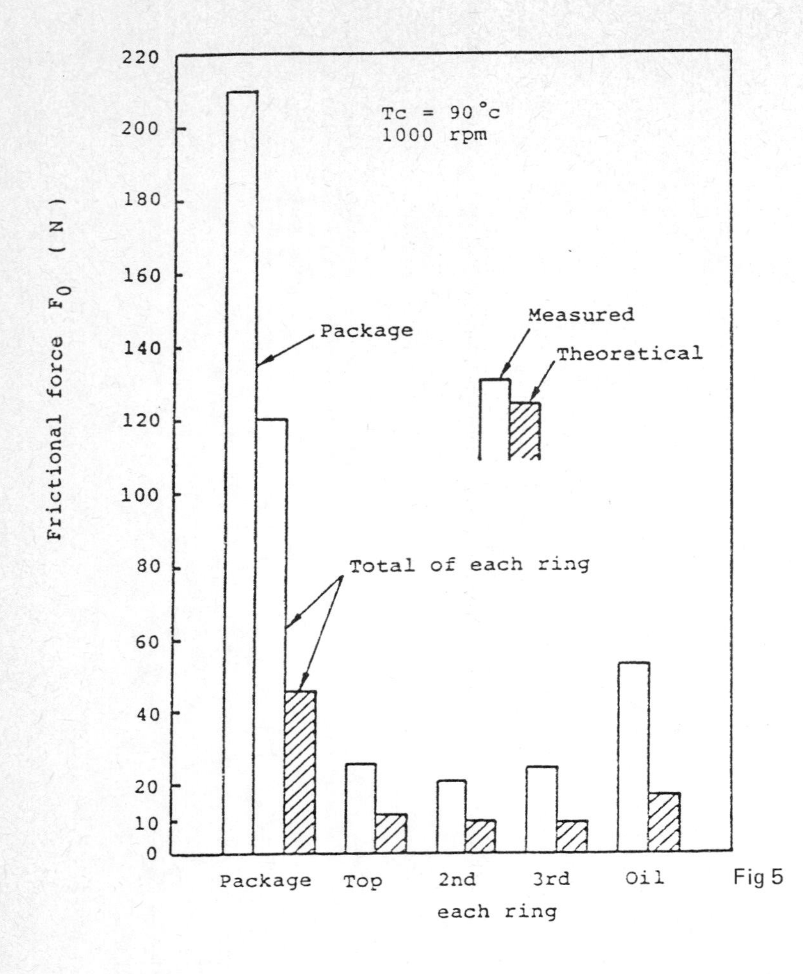

Fig 5 Comparison of measured frictional force of each ring and as a package with theoretical one in case of diesel engine

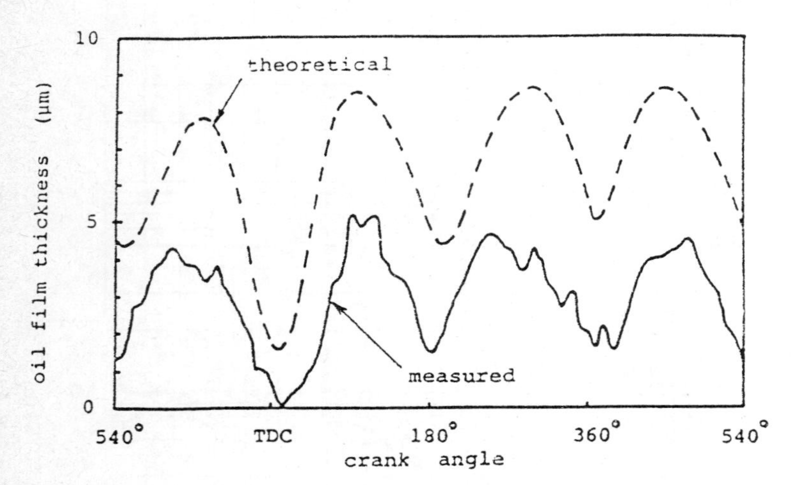

Fig 6a Oil film thickness of top ring in the piston of the standard ring arrangement (four-rings) under full load at 1300 r/min

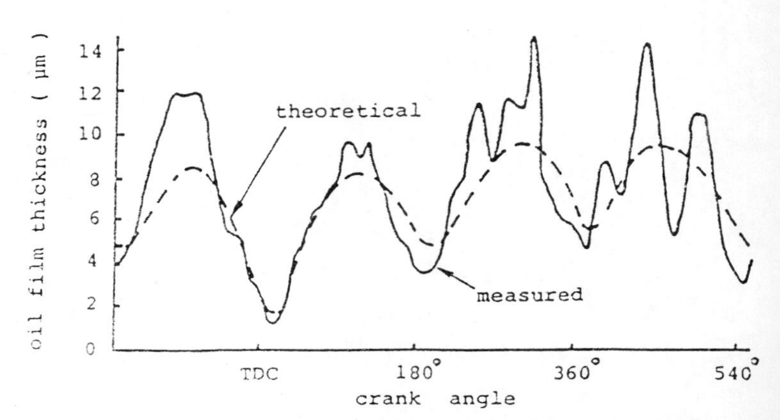

Fig 6b Oil film thickness of top ring when top ring alone is used

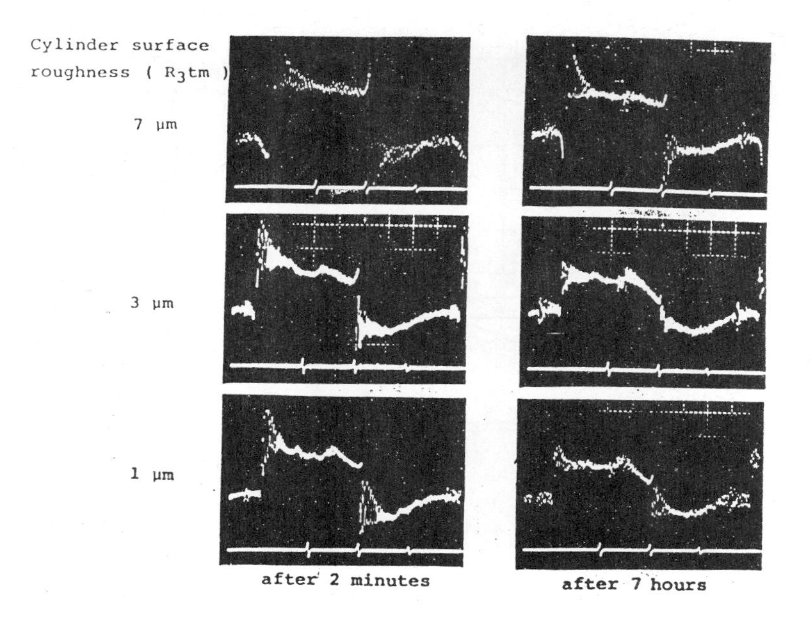

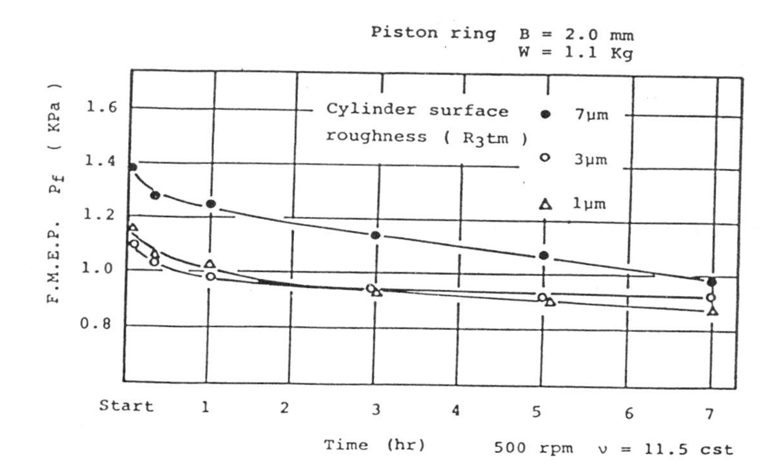

Fig 7 Frictional loss process for running-in with new ring and new cylinder when the cylinder roughness changed

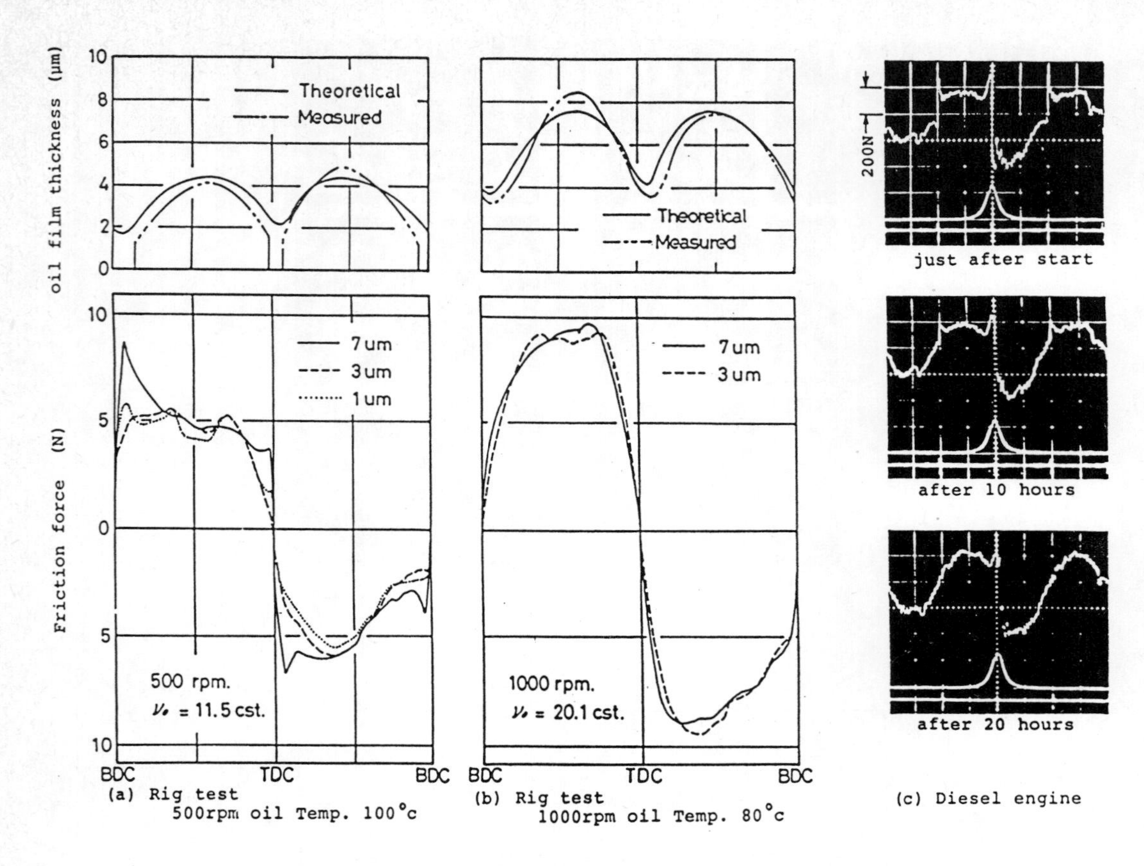

Fig 8 Running-in process starting with new rings and new liner using rig tester and diesel engine

A set of experimental methods for determining the friction and wear behaviour of engine lubricants

J VAERMAN, PhD, DrSc and R De CRAECKER, PhD
Labofina SA, Brussels
B LEDUC, DrIr and G De BRUILLE, Ir
Université Libre de Bruxelles, Belgium

SYNOPSIS An original approach was devised to evaluate accurately the internal friction of combustion engines and its variation with operating conditions and lubricant properties. Two identical SI engines were installed at a fired and a motored test bed. Fired engine parameters at fixed load and speed conditions were duplicated on the motored engine by control of local temperatures and by simulation of cylinder pressures. Neat friction was calculated from the motoring torque and the cylinders pressure diagrams. Correlation of these friction data with fired engine results was ascertained by direct IMEP-BMEP evaluation. Some test results are discussed. A further insight at the relation between friction and cylinder wear was attempted by the use of the visible piston ring technique.

1 INTRODUCTION

The energy crisis has promoted the development of energy conserving techniques. In the field of automotive vehicles, engine and equipment manufacturers have worked to improve the car and engine efficiency. The oil industry has cooperated by formulating fuel efficient engine oils aiming at the reduction of engine friction. Their development and application have been an incitement to have a new and deeper look at basic engine lubrication problems.

Labofina, research center of the Petrofina Group, has developed fuel efficient engine oils and, in cooperation with the Institute of Applied Mechanics of the Free University of Brussels, did experiment some new techniques to evaluate friction in internal combustion engines.

2 EXISTING FRICTION TEST METHODS

Evaluation of the fuel efficiency or of the friction reducing character of an engine oil can be performed by a wide variety of methods ranging from the simulation friction machine, friction measurements in engine sub-assemblies, motored engine tests, fired engine tests, vehicle chassis dynamometer tests, programmed fleet tests, up to the consumer fleet tests. It has been suggested earlier that, with increasing test complexity, up to the service fleet test, the accuracy would decrease due to the difficulty of controlling a great many of variables. On the other hand, the degree of correlation with the practice would show the opposite trend.

The selection of a test method will mainly depend on the objective : certification or development.

Oil certification with respect to fuel efficiency criteria has to be performed along standardized and accepted test methods : in the USA along the ASTM five cars method (1), and in Europe along the Codes of Practice under development in the CEC CL 27 Project Group (2).

Such methods should be able to evaluate fuel consumption improvements of the order of magnitude of 2 to 4 per cent for an oil with respect to the mandatory reference oil. Their accuracy is estimated at about 0.7 per cent in fuel economy.

The complexity and the accuracy level of such tests prevent their use for Research and Development purposes.

Studies of the influence of engine operating conditions and lubricant properties on engine friction require more controlled and accurate methods. The scope of this presentation covers this last type of methods.

We have also undertaken the study of the relations between oil viscometrics, engine friction and wear phenomena although European Car Manufacturers impose a minimum high-temperature-high-shear viscosity limit as a safety rule.

3 OUR ENGINE FRICTION MEASUREMENTS

3.1 Outline of our test methods

A compromise had to be found between accuracy and correlation with service.

In the fired engine tests, one measures fuel consumption for a given power output, but the relation between fuel consumption and power is not only affected by friction and mechanical losses but also, to a larger extent, by thermodynamic losses which are difficult to control.

In a motored engine test, the motoring power is directly equal to its mechanical friction losses plus the pumping losses which are constant for each set of conditions. The initial poor correlation of the motored engine with service has been improved by a properly selected adaptation of its temperature and cylinder pressure levels.

Our basic test approach consisted of two test beds, one motored and one fired, using the same engine model. A high speed data acquisition and computing system was developed and used on the two test beds to measure the average cylinder pressures during the cycle and the internal mean effective pressures.

For these tests, the two liter Renault 829 engine used in the Renault 20 TS car was selected because it was widely adopted by the CEC CL 27 group for their fuel efficient oil test programme as being of modern design and widespread use.

A fired single cylinder Diesel laboratory engine, the Petter AV-B, was also used for friction evaluations (indicated mean effective pressure - brake mean effective pressure) in relation with wear, but this will be dealt with in paragraph 4.

3.2. The motored Renault 20 test bed

The schematic lay-out of this test bed is shown on figure 1. A 22 kW asynchronous electric motor drives the Renault 20 engine by means of a pulley and belt speed variator and a speed and torque sensor. The Renault engine is fitted with a pump driven external oil temperature control circuit and with a glycol coolant temperature control circuit allowing temperatures up to 150°C. Inlet and outlet manifolds are connected by a pressurized and temperature controlled air loop. This device was used to simulate engine load by applying in the cylinders of the motored engine average cycle pressures equal to the average cycle pressures for a fired engine operating under corresponding conditions. This procedure was derived from earlier discussions on the dependence of engine friction on cylinder pressures (3). The temperature of this air loop at the engine inlet was thermostatized at 200°C so as to have closer to service piston and cylinder temperatures and in order to improve test repeatability by avoiding unwanted air temperature variations with engine speed during the tests. The high speed engine data acquisition system consisted of four AVL pressure transducer chains, a crankshaft position sensing chain (pulse generator or encoder) and a Nicolet 1170 averaging oscilloscope.

In this way, the pumping losses of the motored engine could be evaluated by computer integration of the 4 pressure-volume curves, averaged over 128 cycles.

Simultaneously the integral $\oint p \cdot |dV/V_s|$, with V_s being the swept volume, is calculated to obtain the average cylinder pressure level during the cycle.

Neat engine friction data, expressed as FMEP, were obtained as the difference between Motoring MEP derived from the motoring torque and Indicated MEP from the pumping losses as illustrated in figure 2.

Evaluation of the IMEP requires very accurate TDC reference timings (4), (5), (6). We have developed a TDC timing method based on a very precise static positioning of the pulse generator, followed by a dynamic check of synchronization according to (5). The static positioning method uses the encoder signals (4 trains of 2048 pulses per revolution, each with a phase difference of $\frac{1}{4}$ period) to locate the piston of the 1st cylinder at its TDC with a precision of $\frac{1}{4}$ period of $\frac{1}{4} \cdot 360/2048 = 0.044°$ crank angle. This accuracy gives a precision of 0.02 bar for the IMEP in the worst case of the highest circulating air pressure of 1.2 bar effective.

3.3 The fired Renault 20 test bed

This test bed had to fulfil several purposes :

(a) Acquisition of temperature and pressure data as a function of operating conditions, to be applied to the motored Renault 20.

As far as this first point is concerned, it was not possible to perform a full mapping exercise for this engine and it was decided to select a set of operating conditions which may, as far as possible, represent real operation. Three load conditions were selected to be simulated on the motored engine : idling, the average load found in GFC engine operating conditions (2) and full load. For these 3 load conditions, average cylinder pressures of 3, 9 and 18 bar were found and the pressures to be applied in the motored engine air loop were calculated to be respectively -0.2, +0.3 and +1.2 bar effective.

Lubricant and coolant temperatures vary, under real operating conditions, from ambient to operating temperatures. It was not possible to investigate the full temperature range, therefore we selected three temperature conditions. To simulate start-up and low temperature operation (coolant in the fired engine at maximum 45°C), we selected a liner temperature of around 60°C and therefore in the motored engine a coolant temperature of 54°C and an oil temperature of 30°C. These values take in account the cooling performance of our heat exchangers over the load and speed range. For warmed up conditions the GFC procedure requires for the fired engine at low and medium load, 84°C coolant temperature, 95°C oil temperature and 110°C liner temperature, simulated on the motored engine by 99°C coolant temperature and 95°C oil temperature. For warmed up conditions at full load, the fired engine implied 84°C coolant temperature, 105°C oil temperature and 130°C liner temperature. This situation was simulated in the motored engine by 110°C coolant temperature and 105°C oil temperature. This maximum oil temperature may seem somewhat low for full load operation in modern engines. It is however the full load oil temperature derived from the application of the GFC procedure L 007-83 (2) with its particular temperature self controlling system.

(b) Determination of engine friction as IMEP-BMEP.

For direct friction measurements in the

Renault 20 fired engine, the same high speed data acquisition and computing system was used as in the motored engine.

(c) Determination of fuel efficiency of oil formulations.

Finally the Renault 20 fired test bed could be used to evaluate the fuel efficiency performance of engine oils with respect to reference or standardization oils along the French GFC L 007-83 test procedure (2) or a modification of this method. As the repeatability of the results of such fired fuel economy tests is influenced in time by many difficult to control factors such as atmospheric pressure or engine condition, we did develop and experiment a Flying Flush oil change system. This system allows a 100 per cent oil change in 10 minutes without stopping the engine or even changing its operating conditions. The system principle is illustrated in figure 3. Three oil reservoirs, thermostatized at the temperature of the sump outlet of the operating engine, can supply the candidate and two reference oils. On request, a 12 liter volume of the engine oil, equivalent to the extended oil circuit capacity, is drained from the engine sump whilst simultaneously the same volume of experimental oil is pumped from the appropriate reservoir by the engine oil pump in the engine system. This operation is fully automated. After a few minutes rinsing around with the experimental oil charge, a second Flying Flush with the same oil improves the effectiveness of the oil change.

3.4 Present investigation programme

With the two developed test beds, we have performed the following investigations, some of which are still in progress :

(a) Fired engine : measurement of temperatures and average cylinder pressure at preselected representative operating conditions.

(b) Fired engine : fundamental study of FMEP - dependence on speed, load, operating temperature and oil viscosity classes.

(c) Motored engine : simulation of preselected representative operating conditions by an appropriate test bench set-up.

(d) Motored engine : fundamental study of FMEP - dependence on speed, air pressure, operating temperatures and oil viscosity classes; check of simulation value.

(e) Motored engine : first approach of candidate oils by comparison of their motoring torque with that of a reference oil, at the representative conditions.

(f) Fired engine :

- consumption measurements along the official GFC procedure for certification purposes;
- consumption measurements and evaluation of Δ FMEP = Δ BMEP with frequent flying flushes : improvement of the official certification procedure.

3.5 Some experiments and results

It is beyond the scope of this presentation to publish extensively our results, but we will comment somewhat on those aspects which are linked to friction conditions in the engines or to the test procedures themselves.

3.5.1 Effect of load on friction

The effects of oil viscosity and engine speed on friction losses have been widely analyzed and published. The effect of the load parameter on friction in internal combustion engines does not seem so straightforward.

It has been shown (7) that, for bearings operating under hydrodynamic conditions, the load effect on friction is nearly negligible. This seems quite evident insofar as the hydrodynamic lubrication condition is maintained when the load is increased.

For a complete internal combustion engine, the friction response to load will possibly be very complex because there are so many different friction pairs in an engine.

Figure 4 shows some results for the relation between Friction Mean Effective Pressure and load, obtained on the fired Renault 20 engine.

Both at low and high temperature, the trend indicates a decrease in FMEP with increasing load. This seems somewhat surprising but such effects have also been observed, possibly to a smaller extent, in the motored Renault 20 engine. The decrease in engine friction with increasing load will probably be due to an increase of the thermal load and of the oil film temperatures in the cylinders. Friction decreases then with the viscosity of the liner oil films.

If the load increase had induced some engine components to go over from hydrodynamic lubrication to boundary lubrication, one might have expected an increase in friction with load. These experimental results and some others which will be dealt with later, suggest that, for this engine type, most if not all lubrication is hydrodynamic. These results show also the greater importance of the local film temperatures than of the average cylinder pressures.

3.5.2 Correlation between motored and fired friction

The evaluation of friction in an engine is not straightforward and absolute measurements of FMEP in fired or motored engines may prove more difficult than the evaluation of the relative fuel efficiency of two oils in an engine.

Figure 5 shows the degree of correlation between the two test methods for calculated equivalent test conditions of load and temperatures along the speed range : for the fired engine, oil temperature (85 to 102°C) and coolant temperature (84°C) were dictated by the GFC method. The engine torque was 55 Nm. The corresponding values for the motored engine were 95°C as oil temperature, 99°C as coolant temperature and an effective circulating air pressure of 0.3 bar. One will notice that if the FMEP values do not agree exactly, this could be explained at least quantitatively by the differences in the respective liner temperatures that still differentiate the two types of tests with their specially adapted test conditions.

3.5.3. Influence of lubricating conditions on friction

One of the main concerns of car manufacturers about low viscosity fuel efficient engine oils is the fear of lubrication problems with the thinner oils especially when the engine operating temperatures tend to increase. Reduced oil viscosity may increase the occurence of boundary lubrication with the possibility of increase in friction and wear. Friction reducing additives may there be helpful.

Figure 6 shows, over the speed range and for three load and temperature conditions, the difference in FMEP between a viscous SAE 40 reference oil and a lighter SAE 20 newtonian reference oil. This difference is essentially constant over the speed range, but at high loads and especially in the high temperature range it tends to decrease at low speed and even to become negative, indicating thus that friction with the lower viscosity oil increases due to progressive establishment of boundary lubrication. It is however important to note that to show evidence of boundary lubrication in this R20 engine, one has to resort to very severe conditions of low speeds, high loads and low oil viscosity, and of high oil and coolant temperatures. This would show that, in this Renault engine, lubrication is essentially hydrodynamic and that, therefore, the fuel efficient character of the oil will be essentially related to its viscometrics. Here, friction modifiers will not be able to achieve significant improvements. This type of conclusion may however be highly engine dependent as some authors have claimed "fuel efficiency effect obtained by friction modifiers" of the same order of magnitude as what can be obtained by viscometric effects (14).

4 RELATIONS BETWEEN FRICTION AND WEAR

4.1 Wear of internal combustion engines

Although there is no simple relation between engine friction and wear, there may be a strong dependency between these two phenomena, especially in the case of poor lubrication where metal to metal contact can occur.

Wear of the various parts of engines may appear under various forms : adhesive wear, fatigue wear, corrosive wear and abrasive wear.

Wear can also occur at various rates, from the sudden mechanical incident of engine seizure to what could be called accelerated wear such as premature camshaft failure, down to low wear rates such as the accepted normal component wear under the proper operating circumstances.

Wear occurence and rate will depend on a great number of variables, one of which is engine oil viscosity.

A reduction of the oil viscosity below the recommended viscosity grades, in order to reduce engine friction, may increase the rates of the various wear phenomena, possibly by a progressive increase of the frequency of boundary lubrication conditions.

The relationship between oil viscosity and engine wear has been reviewed earlier (8). It was concluded that bearing wear could become a problem but that the influence of a decrease in viscosity on the wear of other components was less evident for two reasons : wear phenomena in valve trains and cylinder assemblies are much more complex from a mechanical point of view, and known effects of lubricating oil additives control and affect the extent of wear, so that viscosity effects may be only of minor importance.

Further work (9),(10),(11) has confirmed the influence of oil viscosity on bearing wear but, as viscosity is not the only influencing factor, there is yet no agreement on an acceptable minimum safe oil viscosity limit. This is suggested to be in the range of a high shear rate dynamic viscosity of 3.5 cP at 150°C down to 2 cP at 150°C.

As far as ring and liner wear is concerned, viscosity effects have been evidenced by radiotracer techniques (12) and more conventional bore measurements (13). This work is however only indicative of a tendency.

As fuel efficient engine oils must perform better than conventional oils during all the engine life span, it was felt that further study of ring and liner wear with special techniques would be interesting.

4.2 The visible ring technique

As shown on figure 7, the gap of the top piston ring of a Petter Diesel engine is observed with a TV camera through a quartz window installed into the cylinder wall at a level corresponding to the bottom dead centre position of the ring.

A stroboscopic flash synchronized with the bottom dead centre provides sufficient light through a semi-mirror.

The piston ring is prevented from rotating by a peg so that the piston ring gap always faces the viewing window. With the optical system, the picture magnitude is increased by a factor 35.

During engine operation, the ring wear induces an increase of the ring gap that can be accurately measured on the TV screen. Increased ring wear rates, devised to reduce testing time and to simulate increased specific ring pressures, can be obtained by using specially manufactured thinner rings.

4.3 Tests on relation between friction and wear

We have tried to investigate the effects of oil viscosity on engine friction and on ring and liner wear and hence, on engine durability. Due to the large domain of load and speed conditions to be examined, a fast wear evaluation method was considered most appropriate.

Tests were performed at constant speed and two load levels with two similar Petter Diesel engines. One of the engines was equipped with the visible ring technique, the other could provide the friction mean effective pressure FMEP by means of a real time data acquisition of the cylinder pressure diagram.

Typical results for the relation between friction and wear for this type of engine are presented on figure 8. A reduction of engine friction, induced by lower oil viscosity, increases the top piston ring wear. The increase in ring wear seems to become very important and possibly unacceptable at full load with oils of too low viscosity.

Further study is required to evaluate properly the compromise which may be required between friction losses reduction and engine wear increase.

5 CONCLUDING REMARKS

A wide set of experimental techniques has been developed to evaluate friction losses in engines, fuel efficiency of lubricating oils and some modes of engine wear. These techniques aim as well at friction-in-engine research as at fuel efficient engine oil development.

Proper adaptation of the operating conditions of a motored engine allows to obtain very accurate friction loss results that are relevant to those of the normal fired engine.

Special attention was devoted to the effect of engine load on engine friction. A slight reduction of overall friction with increasing load was measured and discussed. Some aspects of the load effect on engine wear were investigated in relation with friction.

Tests have evidenced that severe operating conditions may render engine lubrication difficult or even borderline.

Due to the complexity of the internal combustion engine and its wide range of operating conditions, a lot more effort will be required for better comprehension of friction and wear.

ACKNOWLEDGEMENTS

The authors are indebted to IRSIA, "Institut pour l'Encouragement de la Recherche Scientifique dans l'Industrie et l'Agriculture", for a research grant and incitement to perform this study.

REFERENCES

(1) ASTM D-2 Proposal P 101. Proposed method of measuring energy-conserving quality of engine oils in vehicles. ASTM Standards, 1984 vol. 05.03, 1110-37.

(2) GFC Fired Bench Method L 007-83.

(3) MARTIN J. Méthode de mesure des pertes par frottement dans les moteurs à combustion interne. Acta Technica Belgica, EPE, 1971, vol. VII, n° 1, 3.

(4) VAN AKEN Ch. The correct location of top dead centre on indicator diagrams. Revue M, 1972, vol. 18, n° 1, 9.

(5) PINCHON Ph. Calage thermodynamique du point mort haut des moteurs à piston. Revue de l'IFP, 1984, vol. 39, n° 1, 93.

(6) ULBRICH W. Einflüsse von Fehlern bei der Druckverlaufaufnahme auf die Druckverlaufsauswertung. Fisita-Kongress 1984, Wien, nr 845107.

(7) THIELE E. Determination of frictional losses in internal combustion engines. MTZ, 1982, 43, n° 6, 253.

(8) STEWART R.M., SELBY T.W. The relationship between oil viscosity and engine performance, a literature search. SAE 770372 (SAE SP 416).

(9) STAMBOUGH R.L., KOPKO R.J. The relationship of journal bearing wear to multigrade engine oil viscometric properties. SAE 770627 (SAE SP 419).

(10) LONSTRUP T.F., SMITH M.F. Jr. Engine oil and bearing wear. SAE 810330.

(11) VAN OS N., RHODES R.B., COVEY D.F. A study of lubricating oil performance in a journal bearing rig II. SAE 810801.

(12) BELL J.C., VOISEY M.A. Some relationships between the viscometric properties of motor oils and performance in European engines. SAE 770378

(13) PIKE W.C., BANKS F.R., KULIK S. A simple high shear viscometer - Aspects of correlation with engine performance. SAE 780981.

(14) HAVILAND M.L., GOODWIN M.C. Fuel economy improvements with friction modified oils in EPA and road tests. SAE 790945.

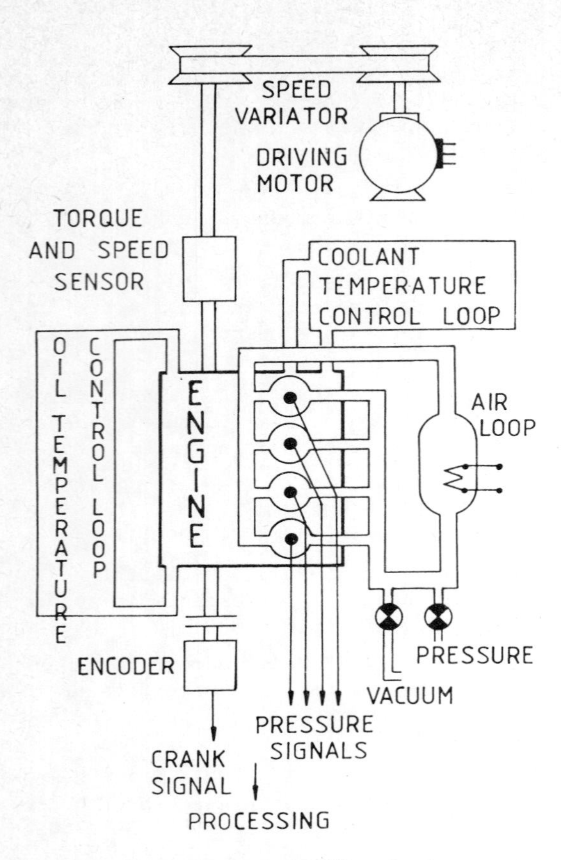

Fig 1 Motored Renault 829 engine layout

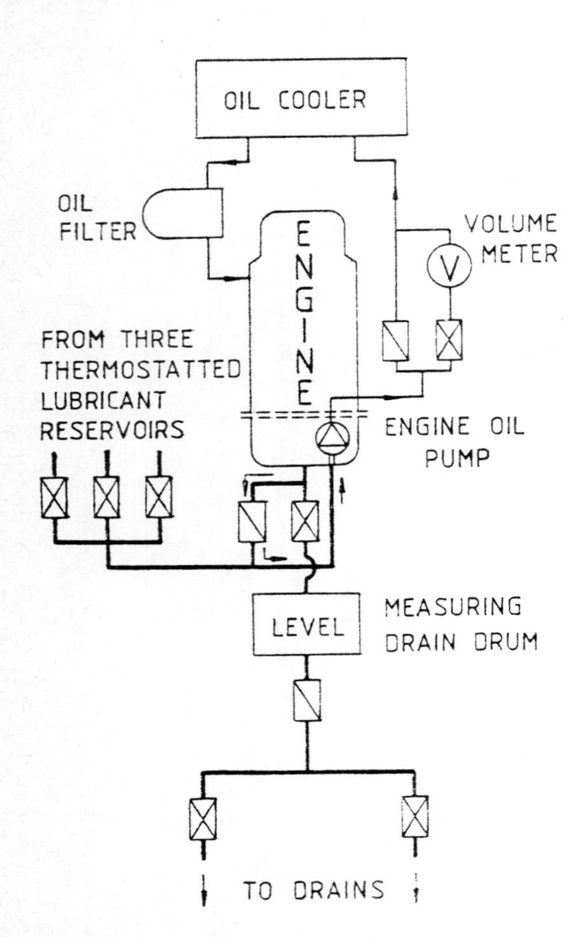

Fig 3 Principle of the flying-flush oil exchange system

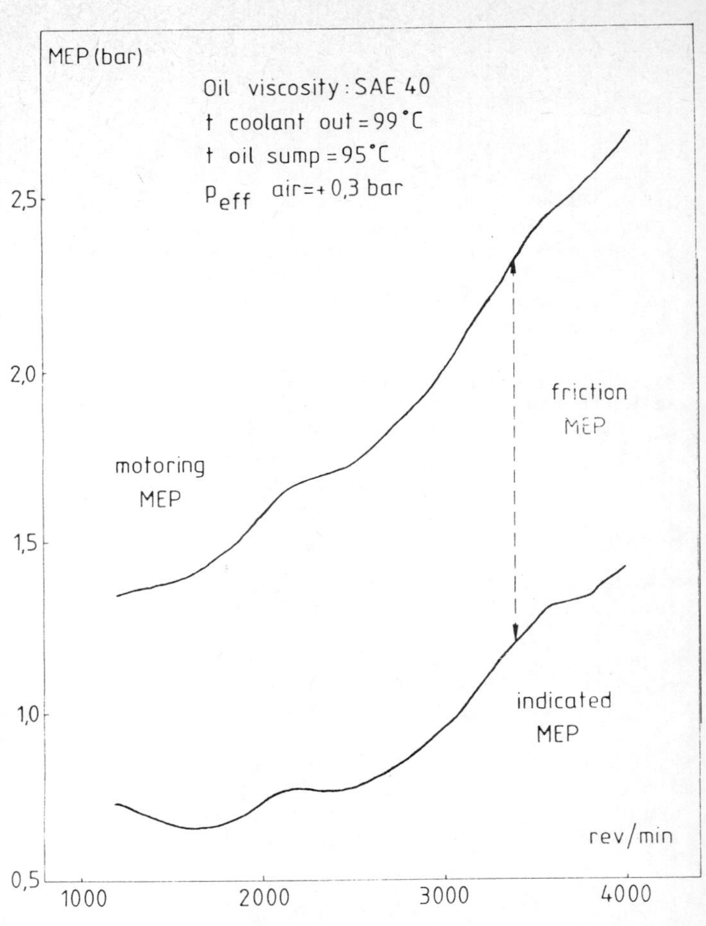

Fig 2 Example of test results from the motored Renault engine

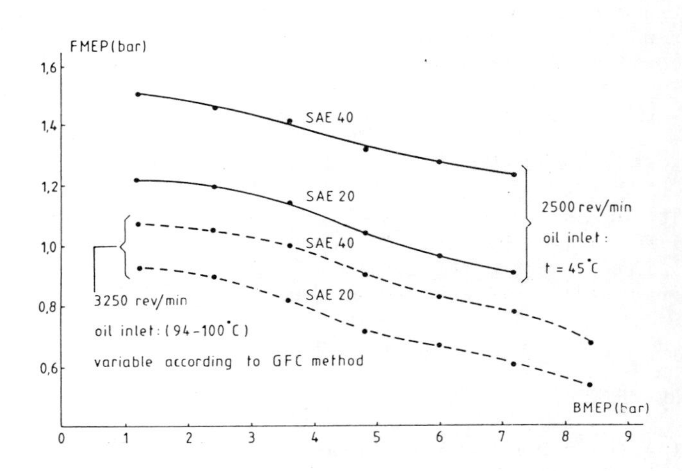

Fig 4 Influence of oil temperature and viscosity grade on the FMEP/load dependence for the Renault fired engine

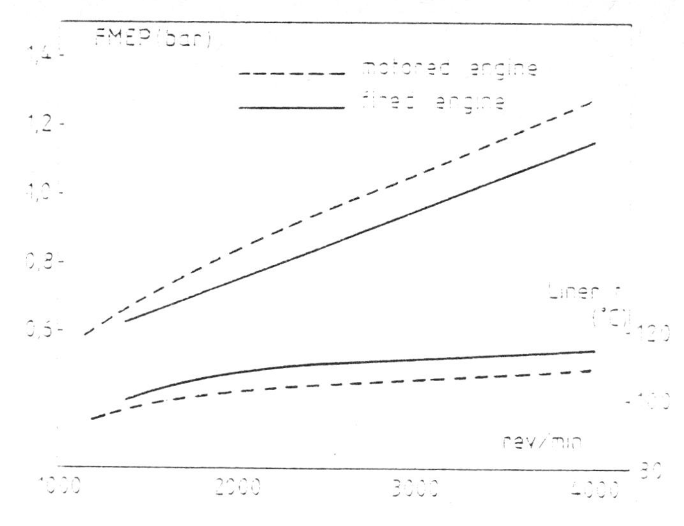

Fig 5 Comparison of the FMEP values and linear temperatures obtained in the warmed-up motored and fired engines for a SAE 40 oil at an average road load

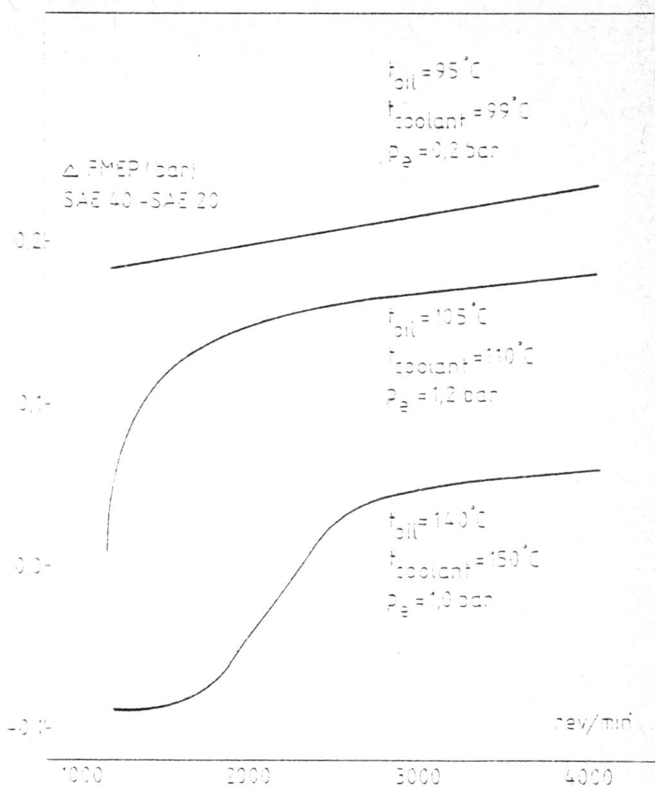

Fig 6 Difference of friction losses between a SAE 40 and a SAE 20 oil under various test conditions (motored engine results)

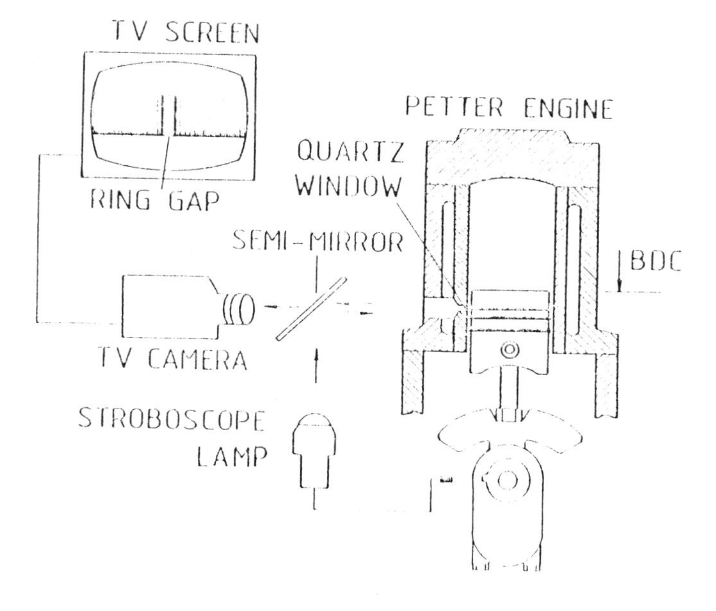

Fig 7 Principle of the visible piston ring technique

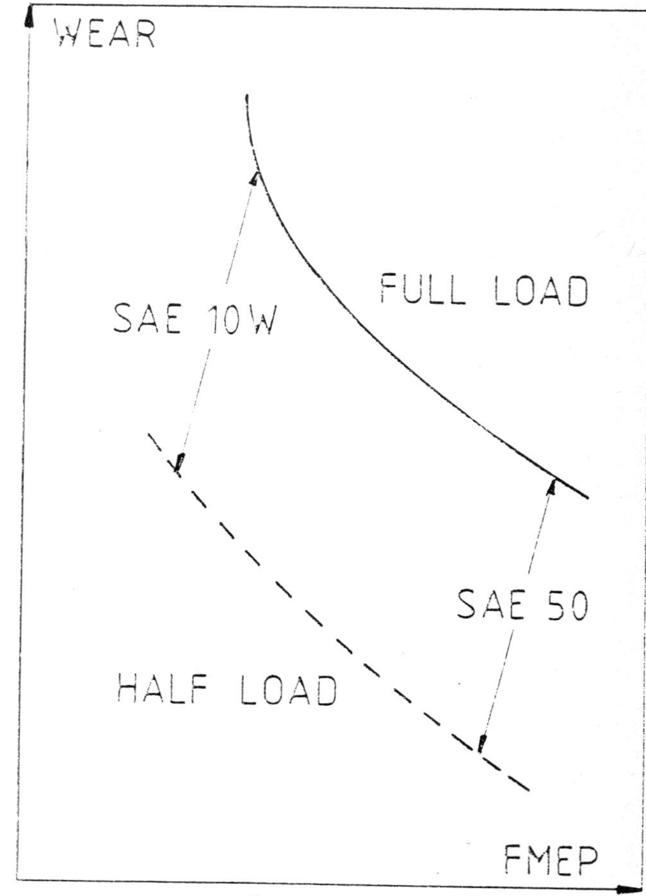

Fig 8 Relations between wear and friction at constant speed and variable load for various oil viscosities

C66/85

Experimental techniques for selecting an engine for fuel economy measurements of lubricants

D J SMITH, BSc and D W GRIFFITHS, BSc, PhD
PARAMINS Technology Division, Esso Chemical Limited, Abingdon, Oxfordshire

SYNOPSIS There is a need for a technique to discriminate between fired bench engines which are being used for the measurement of lubricant derived fuel economy. The basis for this discrimination should address both viscosity and friction modifier effects in conjunction with field correlation.

This paper describes in detail the development of two techniques which have been used at the PARAMINS Laboratories in Abingdon to help in this selection process.

The Morse Test was identified as a suitable technique for mapping engines to evaluate lubricating regimes. However, the original method was not found to be sufficiently accurate and so modifications were made. The theory and development of these modifications together with the experimental results on several different engines are included.

The crucial relationship between the static conditions of a fired bench engine and the transient conditions of normal field use has been considered. Data is presented to illustrate the importance of this relationship.

1 FRICTION REGIMES

1.1 Basic considerations

To obtain a full understanding of the improvements in fuel economy given by lubricating oils it was necessary to know whether the improvements were due to hydrodynamic or boundary friction changes. With this information it would then be possible to produce the next range of fully formulated oils with the optimum viscosity and addition of friction modifiers. Test methods, which operated under known levels of boundary and hydrodynamic friction, needed to be defined.

To achieve this a number of methods were considered. The one most frequently used in the industry is to increase the temperature of an oil at constant speed and load such that the operating viscosity drops. While operating under hydrodynamic conditions the brake specific fuel consumption will decrease with increasing temperature until a point is reached where the lubricant film breaks down. At this point the engine will operate under boundary conditions and the b.s.f.c. will begin to rise with increasing temperature (Figure 1).

Taken in the context of our basic goal of achieving a test procedure this method was not felt practical or relevant. It was considered that we would be forced into a test procedure which would run at different temperatures with the attendant problems of long stabilisation times. Also the test temperatures needed to meet our needs may not be typical of current driving conditions.

Instead, it was decided to fix the oil and coolant temperatures at conditions representative of field performance and to select speed and load conditions to give the variations needed. Using the same basic theory as the previous method two oils of different viscosities were tested and the frictional horsepower of the engine was measured on each over the speed/load envelope of the engine. The well known relationship between friction co-efficient and oil viscosity (Figure 2) then allowed us to determine whether the engine was operating under predominantly hydrodynamic or boundary conditions. If the frictional horsepower increases (i.e. co-efficient of friction increases) when moving from the low to the high viscosity oil then the engine must be predominantly hydrodynamic. If it decreases then it must be predominantly boundary.

Using this method it is possible to establish the contribution of boundary and hydrodynamic friction to the total engine friction at any given speed/load condition and at known fixed oil and coolant temperatures.

To achieve such data, however, it is necessary to measure the friction horsepower of the engine under the required conditions. Many methods have been developed for this purpose but none was totally acceptable in meeting our needs. To meet the aims of our long range plan of surveying a number of engines it was necessary to carry out each test quickly, easily and without high costs.

1.2 Friction tests

Many modifications have been made to the basic Morse test, both in terms of equipment and theory. However, the inherent problems of loss of manifold efficiency and unstable ignition following the misfiring cylinder have yet to be solved satisfactorily.

The use of a motoring dynamometer was rejected for two reasons. Firstly, the engine was not subjected to normal combustion pressures and temperatures. This was felt to be particularly critical to the ring/liner friction where the lubricant film was exposed to exhaust gas temperatures. The practicality of either finding engines already fitted to motoring dynamometers or fitting engines of different configurations to a single bed was also considered to be a problem.

While being the most accurate method available, the use of piezo-electric transducers to measure instantaneous gas pressure and hence indicated horsepower was also thought to be too time consuming and costly.

The test method having the most scope for achieving our aims was considered to be the Morse Test and work was put in hand to try to improve the precision of the test.

1.3 The Morse test

The basic test method involves measuring the power developed by the engine at a given speed and throttle opening and then measuring the power produced at the same speed and throttle opening when each of the cylinders are misfired in turn. This operation produces two simultaneous equations with two unknowns: indicated horsepower and frictional horsepower.

Viz.

Brake horsepower produced where all cylinders are firing

$$= B4 = I1 + I2 + I3 + I4 - F \quad \text{(4 cylinder engine)} \qquad (a)$$

where I# is the indicated horsepower from cylinder # and F is the frictional horsepower

Brake horsepower produced when cylinder 1 is misfiring

$$= B3^1 = I2 + I3 + I4 - F \qquad (b)$$

Similarly for cylinders 2, 3 and 4 misfiring

$$= B3^2 = I1 + I3 + I4 - F \qquad (c)$$
$$= B3^3 = I1 + I2 + I4 - F \qquad (d)$$
$$= B3^4 = I1 + I2 + I3 - F \qquad (e)$$

Hence

$$B3 = 1/4\,[B3^1 + B3^2 + B3^3 + B3^4]$$
$$= 3/4\,[I1 + I2 + I3 + I4] - F \qquad (f)$$

Substituting for (a) in (f)

$$B3 = 3/4\,[B4 + F] - F$$
$$\text{or } \underline{F = 3B4 - 4B3}$$

This simple and neat relationship has been used extensively for a number of years but, due to the operating scheme of the engine it has a major drawback. For fully efficient operation of the engine, it relies on pressure pulses in the inlet and exhaust tracts to be synchronised. If this does not occur then the power is reduced [loss of manifold efficiency]. When a cylinder is not firing the change in pressure pulses affects the power obtained from the remaining three cylinders. Hence equations (b) to (e) should be of the form

$$B3^2 = I1 + I3 + I4 - F - M$$

This then produces two simultaneous equations with three unknowns, which is insoluble.

Modifications to the method have attempted to overcome this by rotating the misfire around the engine on consecutive cycles. This can to some extent overcome the problem but introduces another variable.

The pattern of misfiring cylinders around the engine is as shown in Table 1, and it is apparent that any individual cylinder has three normal firing cycles, one misfiring cycle and one cycle where the firing cycle follows a misfire. This latter cycle produces a different power output from a normal firing cycle due to an excess of fuel/air mixture in the combustion chamber resulting from the non-ignition of the previous cycle. Consequently equations (b) to (e) should be of the form

$$B3^2 = I1 + I3 + I4 - F + P$$

This again leads to two equations with three unknowns.

TABLE 1

FIRING SEQUENCE - STANDARD MORSE TEST

Cycle No.	Cylinder No.			
	1	3	4	2
1	M	F	F	F
2	F	M	F	F
3	F	F	M	F
4	F	F	F	M
5	F	F	F	F
6	M	F	F	F
7	F	M	F	F

1.4 Modified Morse test

To overcome this problem we have developed a multiple function rotating misfire unit. This unit allows the engine to be operated under any of the following conditions.

(i) All cylinders firing.
(ii) One cylinder misfiring with misfiring cylinder rotating around the engine.
(iii) Two cylinders misfiring with no spaces between misfiring cylinders and the misfiring cylinders rotating around the engine.
(iv) As (iii) but with one firing cylinder between the two misfiring cylinders.
(v) As (iii) but with two firing cylinders between the two misfiring cylinders.
(vi) The unit will accept engines having between three and eight cylinders.

For the four cylinder application the firing patterns for the various functions are shown in Table 2. There are four possible power outputs for each cylinder.

I = power under normal running conditions.
IM = power developed when firing cylinder misfired on previous cycle.
IV = power developed when firing cylinder follows a misfiring cylinder.
IMV = power developed when firing cylinder misfired on previous cycle and when it follows a misfiring cylinder.

If the cylinder misfires then no power is developed.

Using Table 3, which shows the power outputs corresponding to Table 2, four simultaneous equations can be produced thus:

$$B4 = 5I - F \quad (i)$$
$$B3 = 2I + IM + IV - F \quad (ii)$$
$$B20 = I + IM + IV - F \quad (iii)$$
$$B21 = IM + IV + IMV - F \quad (iv)$$

with four unknowns viz. I, F, IMV and (IM + IV). Solving these equations produces:

$$\underline{I = B3 - B20}$$

$$\underline{F = 5\,(B3 - B20) - B4}$$

$$B20 = B3 - B20 + IM + IV - (5\,(B3 - B20) - B4)$$

$$B21 = 3B20 + 4B3 - B4 + IMV - F$$

$$\underline{IMV = B21 - 2B20 + B3}$$

(IM + IV) can also be calculated but individual values of IM and IV cannot be separated without further assumptions.

TABLE 2

FIRING SEQUENCE - MODIFIED MORSE TEST

Condition B4

1	3	4	2
F	F	F	F
F	F	F	F
F	F	F	F
F	F	F	F
F	F	F	F

Condition B3

1	3	4	2
M	F	F	F
F	M	F	F
F	F	M	F
F	F	F	M
F	F	F	F

Condition B20

M	M	F	F
F	M	M	F
F	F	M	M
F	F	F	M
M	F	F	F

Condition B21

M	F	M	F
F	M	F	M
F	F	M	F
M	F	F	M
F	M	F	F

TABLE 3

POWER OUTPUTS - MODIFIED MORSE TEST

Condition B4

1	3	4	2
I	I	I	I
I	I	I	I
I	I	I	I
I	I	I	I
I	I	I	I

Condition B3

1	3	4	2
M	IV	I	I
IM	M	IV	I
I	IM	M	IV
I	I	IM	M
IV	I	I	IM

Condition B20

M	M	IV	I
IM	M	M	IV
I	IM	M	M
IV	I	IM	M
M	IV	I	IM

Condition B21

M	IMV	M	IV
IM	M	IMV	M
IV	IM	M	IMV
M	IV	IM	M
IMV	M	IV	IM

1.5 Test results

Using this technique to measure the frictional horsepower, comparisons were made of the engines running on two oils of different viscosity grades (SAE 30 and SAE 10) over the complete speed/load envelope.

Seven engines have been mapped to date (including one air cooled).

Engine size has varied from 1.1 litres to 5.7 litres and covers a range of British, French, Italian and U.S. engines.

Typical friction regime maps are presented in Figure 3. The maps show the areas of the speed/load envelope which are predominantly hydrodynamic or predominantly boundary lubrication (i.e. greater than 50%). It is possible to generate a full contour map by showing the change in f.h.p. as a percentage of the total friction but this has been excluded for ease of presentation.

In addition to the primary requirement, it is believed that many other things can be learnt about the frictional conditions in engines from these maps.

1.6 Review of results

The maps show a wide variation in frictional regimes from engine type to engine type. Type A engine for example is seen to operate under hydrodynamic lubrication for the majority of the area. On the other hand Type B is seen to operate under boundary conditions for a large proportion of the time.

The selection of an engine for fuel economy testing is consequently very dependent on the form of these friction regime maps. An engine such as Type A will in general not show the same level of fuel economy benefits from chemical friction modifiers as would Type B. Conversely Type A will show relatively large improvements for reductions in the operating viscosity of the lubricant. Type B may even show debits in fuel economy for reductions in viscosity.

2 FIELD CORRELATION

2.1 Basic considerations

An engine installed in a vehicle operating in the field is subjected to a wide range of test conditions and driving styles. These combinations of conditions and styles will have a significant influence on operating conditions of the engine. Consequently the same vehicle running under different conditions and styles will have different responses to lubricant formulation.

In an attempt to characterise these effects a test was developed for use in a car.

2.2 Field condition maps

To identify the operating conditions of an engine under various field conditions two sensors were installed on a car engine. The first monitored the engine speed and the second monitored the inlet manifold vacuum, which was used as an indirect measurement of load. These were connected to a data logger which recorded both values every two seconds and transferred them to magnetic tape. The car was then taken out on the road and operated under the following conditions/ styles.

(i) Highway/constant speed.
(ii) Cross country/high speed.
(iii) Cross country/ low speed.
(iv) Urban.

Approximately four thousand data points were collected for each test condition and these were analysed as a contour plot with axes of speed, manifold vacuum (load), no. of records as a percentage. Some typical plots are shown in Figure 4.

2.3 Review of results

The three plots presented; corresponding to conditions (i), (ii) and (iii) above; illustrate the different operating conditions the engine is subjected to.

Figure 4A (highway) shows that the engine spends the majority of its time at a single speed and load with only slight deviations depending on gradient and other traffic.

Figure 4B (cross country/high speed) shows a more variable pattern of use with four distinct areas. Area A is idle. Area B is the steady state road load and areas C and D correspond to wide-open-throttle accelerating conditions in fourth gear (C) and third gear (D).

Figure 4C (cross country/low speed) shows an even more variable pattern having no significant peaks. This indicates the use of all gears and all throttle positions.

From these plots it is possible to choose the speed/load conditions which are most representative of specific driving patterns and to weight these conditions proportional to their frequency of occurrence. Further, the chosen conditions can then be superimposed upon the friction regime maps to establish for what proportion of time the engine is operating under boundary and hydrodynamic conditions.

3 USE OF THE TECHNIQUES

The work carried out to date by PARAMINS has concentrated on reviewing a range of modern production engines to establish a "typical and representative" tool with which to work. Initial results have been encouraging and have shown a significantly larger proportion of boundary lubrication in current engines than first thought; with up to 40% of the working envelope operating in the predominantly boundary regime.

Work will continue to identify a typical engine for use in our in-house program of fuel economy evaluation using the modified Morse test. Test conditions will then be chosen based on the field correlation work to enable the fuel economy measured on the test bed under steady state conditions to be related to the transient conditions of field usage.

4 SUMMARY

The two techniques described in this paper can, if used together, provide detailed information on the changes in lubricant formulation needed to achieve fuel economy benefits in the field.

Used individually the modified Morse test enables engines to be evaluated in terms of their ability to discriminate between lubricants of different technology biases. The basic technique also provides a rapid and more accurate means of measuring frictional horsepower and mechanical efficiency than the original test.

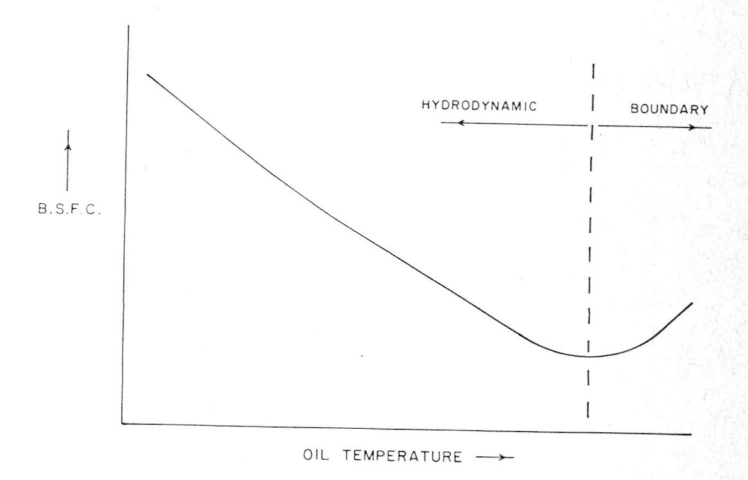

Fig 1

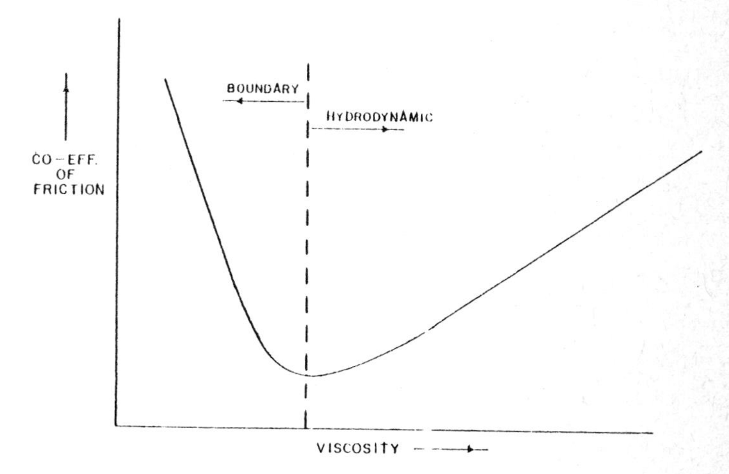

Fig 2

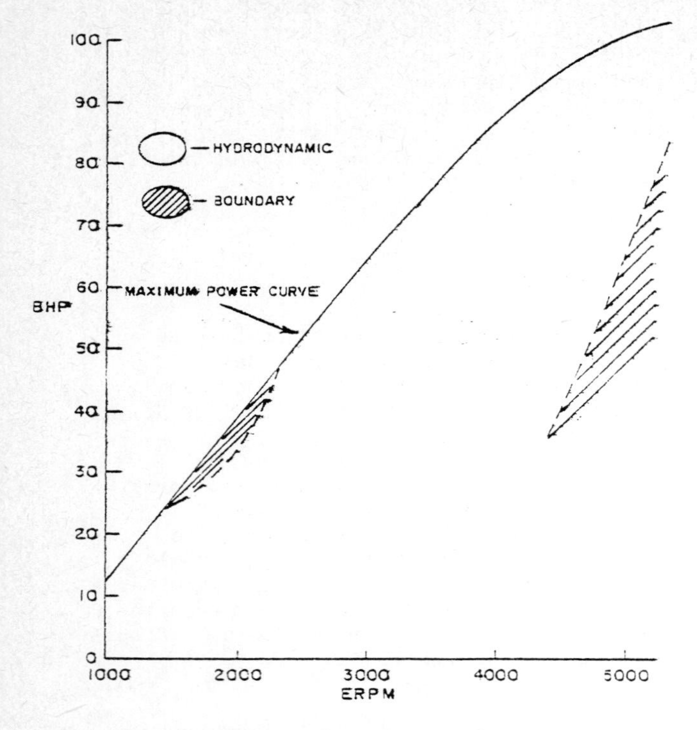

Fig 3a Friction regime map — engine type A

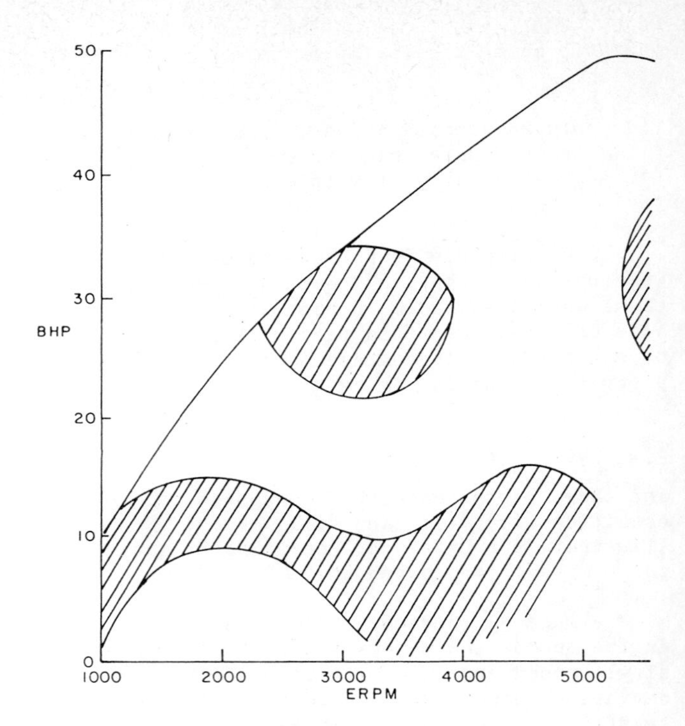

Fig 3b Friction regime map -- engine type B

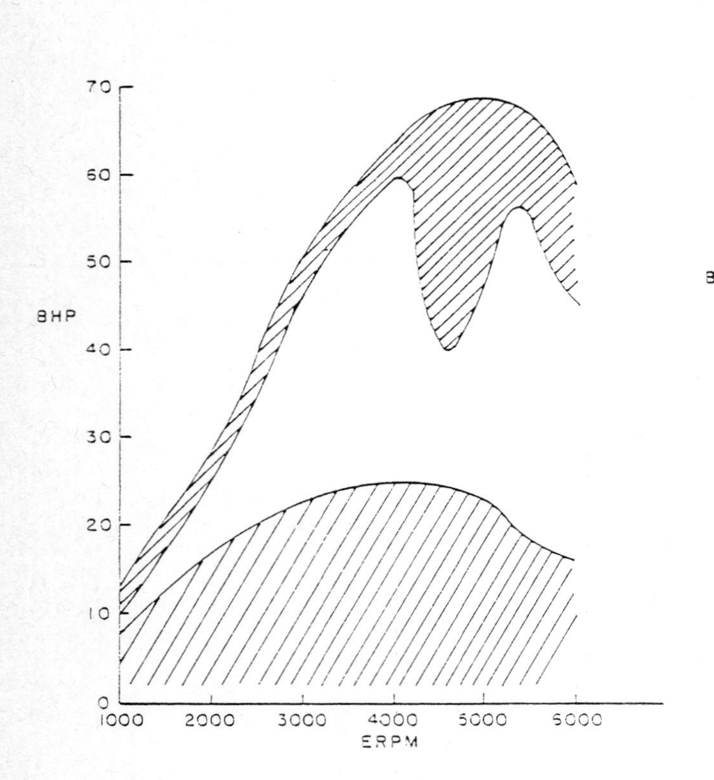

Fig 3c Friction regime map — engine type C

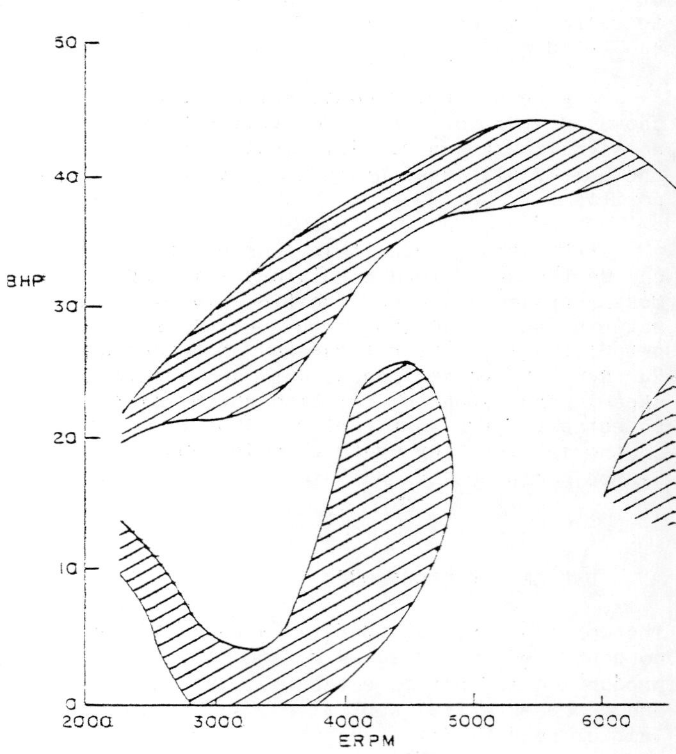

Fig 3d Friction regime map — engine type D

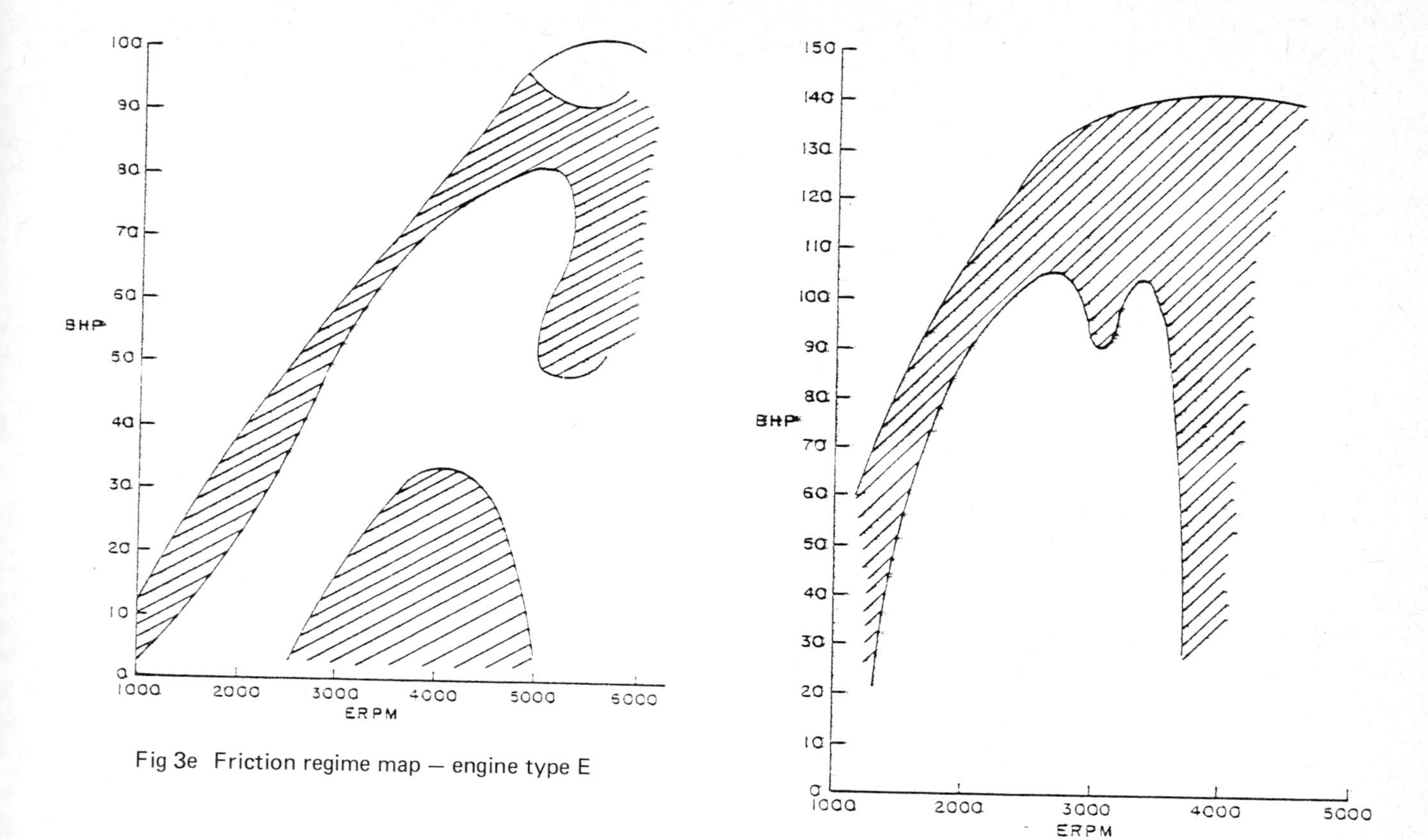

Fig 3e Friction regime map — engine type E

Fig 3f Friction regime map — engine type F

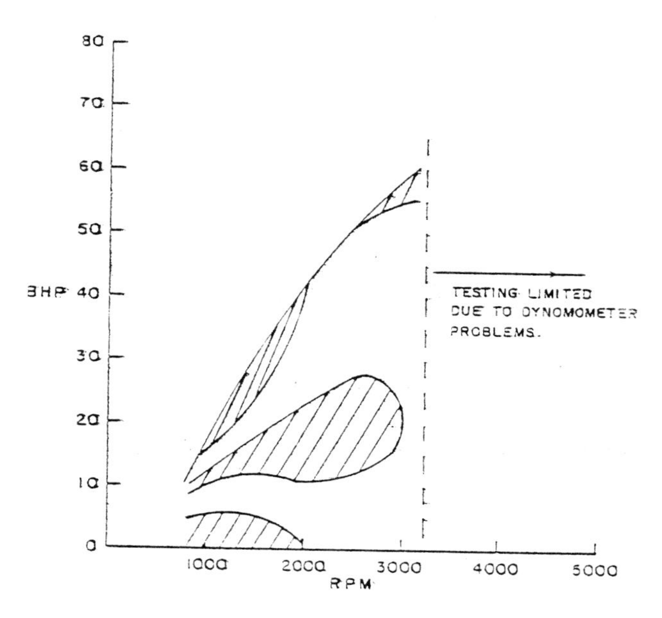

Fig 3g Friction regime map — engine type G

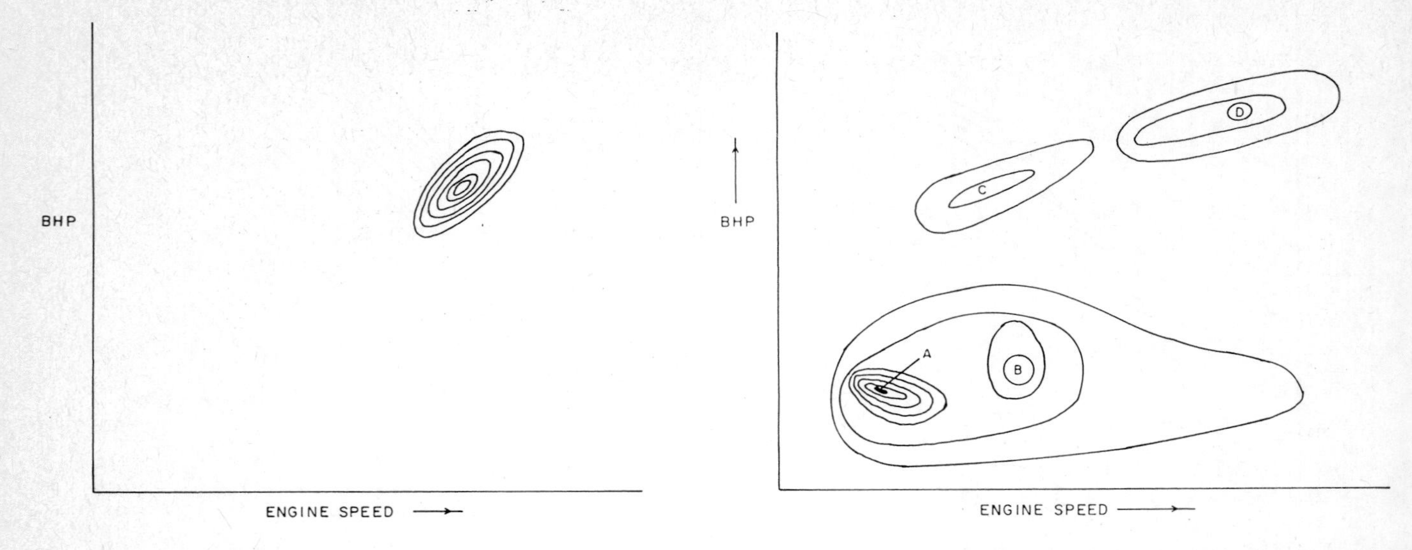

Fig 4a Power/speed frequency plot — highway

Fig 4b Power/speed frequency plot — cross country high speed

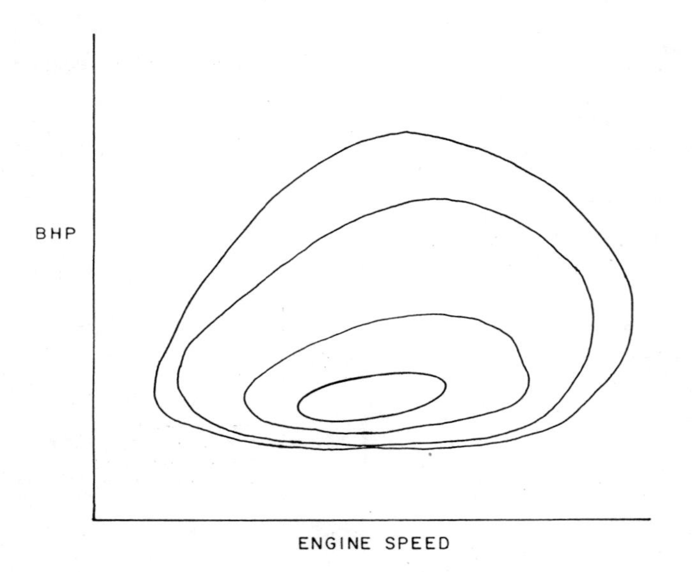

Fig 4c Power/speed frequency plot — cross country low speed

An imaging system for the radioactive tracing of lubricants in automotive components

J HERITAGE, BSc, **T J HOYES**, BSc, Castrol Limited, Pangbourne, Berkshire
P A E STEWART, MSc, CEng, FIMechE, **R G WITCOMB**, BSc, DPhil, Rolls-Royce Limited, Bristol
J E BATEMAN, BSc, PhD, **J F CONNOLLY**, BSc, **R STEPHENSON**, BSc, Rutherford Appleton Laboratory, Chilton, Oxfordshire
P FOWLES, BA, PhD, **M R HAWKESWORTH**, BSc, MSc, PhD, **M A O'DWYER**, BSc, MSc, PhD, and
J WALKER, BSc, PhD, DSc, Department of Physics Radiation Centre, University of Birmingham

SYNOPSIS

One of the shortcomings of using neutron radiography as a means of locating hydrogenous materials within a metal structure is its inability to show movement within, behind or in front of the overall shadow of hydrogenous material. Birmingham University, Rolls Royce and Castrol have collaborated with the Rutherford Appleton Laboratory to develop a completely new system to over come these limitations - Positron Emission Tomography.

The paper outlines the mechanical and electronic design of the positron "camera" and associated computing facilities used to produce both pictorial and quantitative results through increasing thicknesses of metals. It also deals with the lubricant details needed to allow visualisation of various features and their engineering relevance.

1. INTRODUCTION

Lubricant researchers and engine manufacturers already use many techniques to investigate the detailed operation of lubrication systems. The wish to monitor the presence of complex flow patterns and rapid movements of fluid within its bulk has led to our concentration on novel and innovative inspection techniques rather than the more familiar endoscopes, pressure and flow sensors, and windows, which are often too limited in scope or unacceptably intrusive.

X-Rays

The attenuation of x-rays is governed by their interaction with the electrons in the target body and material opacity increases more-or-less smoothly with density. Accordingly it is difficult to produce satisfactory contrast in images of fluid overlaid by metal and only in a few areas where artificial doping of the fluid does not affect its physical and chemical properties is the technique viable.

Neutrons

In contrast with x-rays, neutrons interact directly with nuclei with the result that the relative stopping power of materials is not a smoothly rising function of their atomic weight or density. For example hydrogen and hydrogenous materials such as water, oil or plastic have a high opacity, whereas aluminium (and to a lesser extent iron) do not. Thus neutron illumination of a metal-lubricant system will produce an intensity shadowgraph (in much the same way as x-rays) which can be visualised either statically using a sensitive film or dynamically using a scintillator (a 'radiation to light' converter) and video camera system. Neutron radiography in this context involves the use of a highly collimated beam of low energy neutrons, usually generated within a nuclear reactor.

The technique has been used successfully to monitor lubricant flow within both aero and automotive engines and gearboxes (Refs. 1 and 2) but a limitation is its inability to image behind or in front of an overall shadow. Separate views through the object under investigation may be satisfactory for some applications, but where conditions change subtly and unrepeatably information will not be coherent. Similarly the movement of fluid within its own bulk remains within the single shadow thus denying information on, say, possible stagnant regions within components

2. POSITRON EMISSION TOMOGRAPHY

Within the last ten years a new diagnostic technique for investigating certain medical conditions has been developed. Originally applied to the measurement of brain metabolism it involves the use of certain radioactively 'tagged' compounds which are taken up by various regions of the brain to an extent dependent on their metabolic activity. The technique, Positron Emission Tomography, or PET, permits the measurement of radioactivity in predefined planes within the target so indicating specific regions of cerebral activity.

Early systems were relatively slow (to produce results), heavy (untransportable), expensive and, with relatively poor resolution, were unlikely to extend their application beyond limited areas of medical research. However, developments in detectors and microprocessor technology have now made possible the construction of systems with improved performance and of a design which is truly transportable. Such a system has been constructed by the SERC Rutherford Appleton Laboratories and has visited various hospital sites in the United Kingdom (Ref.3). The design principles used in this new type of camera system were the basis for the larger and more complex unit constructed by RAL for flow studies in industry.

2.1 Principle of Operation

An essential feature of Positron Emission Tomography is incorporation of a positron emitting isotope into the fluid to be monitored. The method of addition can be bulk mixing, or continuous or even pulsed injection, depending on the type of test to be conducted. Positrons are spontaneously emitted from the isotope atoms at an average rate dependent on the half life of the particular isotope. The distance travelled by positrons is dependent on the surrounding material density and in water or oil is typically two to four millimeters. Annihilation with an electron subsequently occurs. The resulting gamma ray photons produced by the annihilation leave in almost exactly opposite directions (Fig. 1).

In the present industrial system two rectangular detectors which intercept and stop some of these photons are placed on opposite sides of the subject. The design and construction of the detectors is such that the coordinate positions at which the photons strike can be read out. High speed electronics ensure that only events occuring simultaneously in each detector and therefore associated with the same positron decay are passed on to the recording system, with many such data records making up a test run.

Image reconstruction involves the use of a powerful computing system which for each data record uses the detection coordinates (and detector separation) to calculate the flight locus of each gamma ray pair and the points at which it intercepts chosen planes within the space between the detector plates. In this way a three-dimensional view of the distribution of radioisotope within the target body can be constructed, and fluid position and quantity identified.

It will be realised that the actual origin of a positron annihilation cannot be identified, only the flight locus of the gamma rays subsequently produced. One event will therefore contribute to many if not all of the image planes generated. The reconstruction process can be usefully clarified by comparing it to focussing an optical microscope on an object within a transparent specimen. For that particular plane there will still be contributions from light from all the other planes within the specimen. However, unlike the object plane, these will be out of focus and their contribution to the overall image reduced (those nearest to the focused plane contributing more than those at a greater distance). Three dimensional imaging is thus achieved.

3. DETECTOR DESIGN AND CONSTRUCTION

The two gamma ray detectors function on the same principles as the smaller medical PET units (Ref. 3). However, some improvements to the design have been made to produce a more robust construction with greatly reduced sensitivity to acoustic excitation, of great importance if the camera is to be used in industrial environments. The camera system can be seen in Fig. 2. Each detector case is 1500mm high by 1750mm long by 800mm deep and with its stand weighs approx. 350 kg. The detection efficiency of each (to incident 511 keV photons) is approximately 10 per cent but as both detectors must be triggered for a true decay to be registered the combined efficiency is approximately one per cent.

3.1 Detector Construction and Operating Principles

The detectors utilise modern Multiwire Proportional Counter (MWPC) technology. Each comprises a stack of 20 MWPC sections with an active area of 600mm by 300mm, giving a total thickness of 124mm. These are sealed in an aluminium case containing isobutane at atmospheric pressure. A section consists of an anode plane of 300 wires each of 20 microns diameter stretched across a glass reinforced plastic frame, and a printed circuit cathode which has been electroplated with lead to provide a converter medium. A high voltage, typically 3.5 kilovolts, is maintained between the anodes and cathodes. A gamma ray entering through the front (thin aluminium) window has a certain probability, dependent on lead thickness, of being stopped in one of the cathodes. Energy transfer gives rise to emission of an electron which produces an ionisation trail as it travels through the surrounding gas. The high voltage gradient near the thin anode wire produces a controlled 'avalanche' of electrons which is detected by the anode electronics. This pulse is also sensed by the two nearest cathode planes. Positional information is achieved by having the cathode coating divided into 3mm strips, with alternate cathodes having strips running orthogonally. Each strip is fed into an electronic delay line running the length of each detector, and the difference in time of arrival of the anode pulse, and cathode pulses at the end of each delay line is used to calculate the X & Y coordinates of photon interaction.

3.2 Event Identification & Recording

To ensure that only true decay events are recorded, acceptance depends on the two separate detectors being triggered within a 20 nanosecond window of each other. This avoids two random single signals producing false records. When a positive coincidence is found it provides a basic trigger for the complete readout system. Other checks are also conducted on the detector data to ensure a valid event is stored. Assuming all checks are positive, coordinates from each detector are sent via an interface system to a mini-computer running in a logging programme mode for subsequent hard storage on magnetic tape or removable hard disc, from which it can be later retrieved and processed. Decay events can be processed and stored at rates up to several kiloHertz, and 1.7 million events are produced in a typical test run (the maximum capacity of one disc).

A further valuable feature is the incorporation of an analogue computer and display which produces a simple image for one plane only (user selectable) within the object space. This provides a useful facility for initial preparation and diagnostic work.

4. COMPUTING

Considerable computing power is utilised in both areas of system operation. These are:

1) control of camera, data formatting and recording during a run

2) data retrieval, calculation of images (back projection), image processing and object overlay (The latter can greatly assist in the interpretation of results).

Two separate processors are used(both mini-computer systems) so that test and analysis work can be conducted simultaneously. For comprehensive analysis, processing is carried out on a main frame system. This may be essential for some of the more complex imaging options under development.

Operation of the logging system is straightforward, being controlled via a suite of 'user friendly' programmes. Pertinent information - test description and detector separation - is also recorded for future identification and calculations.

With data accumulation initiated, each valid event passes via the interface system to the logging computer to be permanently recorded by a high speed tape drive. The process can be interrupted at any stage or left until a full record (of 1.7 million events) is achieved. Owing to the decay of incorporated isotope a compromise between total recorded events and run time is often accepted. Processing is conducted using a further suite of programmes:

a) generation of image planes using the raw data,

b) use of contrast enhancement, or filtering techniques to enhance image features,

c) results interpretation using Digital Object Space Representation (DOSR)

a) Generation of Planes

The distribution of isotope is displayed within selected (up to 64) planes within the target body. The active area of the detectors is divided into 256 x 128 pixels giving a fundamental resolution of ~2.3mm. The basic programme will, for each recorded event, calculate the flight path of the annihilation photons and register the pixels it transfixes in each plane of interest. The process is repeated for each event and the 'hits', or events, in each pixel recorded. With all data processed the number of events (representing isotope activity) can be visualised using a 256-level grey scale, or by assigning colours to particular levels.

b) Image Enhancement

An Intellect 100 image processing system offers several image enhancement options. "Graininess" can be reduced by utilising Fourier filtering techniques. Similarly slight variations in activity or intensity may be emphasised by stretching the 'contrast' over the particular range. Other standard techniques available include area measurement and edge enhancement.

c) Image Interpretation

An additional computer programme known as Digital Object Space Representation (or DOSR) can be used in conjunction with the image generation programme to inform the computer about the distribution of materials around the isotope-doped fluid. This consists essentially of a three-dimensional map of the space between the detectors built up of discrete small cubes or 'voxels', which may be coded according to the type of material they represent. For a particular target a three dimensional model can be generated using computer-aided design techniques (or from the design drawings). The relevant target planes with their intensity information may now be overlaid with the relevant DOSR slice and the presence or absence of doped fluid within oilways, drillings channels etc. may be seen at a glance.

The above programmes and facilities give a pictorial representation of test results. For more detailed, or quantitative analysis the image information can be made available in numerical form and may be extracted and manipulated for quantitative interpretation.

5. ISOTOPE TRACER REQUIREMENTS

The operation of the PET system requires the incorporation of a suitable positron emitting isotope in the fluid whose movement and distribution is to be monitored. Use of the technique for non-medical applications relaxes considerably the restriction on activity levels. Greater quantities can be utilised to permit faster imaging through the denser materials experienced in industrial applications. Whilst positron emitters can be found throughout the Periodic Table of elements, ultimate selection must meet the following conditions and restrictions if successful imaging is to be achieved:

1. sufficient activity to yield useful images over a reasonable time scale,

2. solubility in a wide range of fluids used in equipment that may be subjected to investigation,

3. compatibility with fluids or additives in fluids so avoiding

   a) precipitation of isotope resulting in poor imaging and artefacts;

   b) interference with the correct running of the machine (and possible damage),

4. sufficiently short half life (ideally one minute to one hour) so that after testing the equipment can be quickly returned to service, or prepared for additional investigation,

5. the complete isotope system should be simple enough to be operated by non-experts.

The isotope predominantly used for early proving and evaluation work is gallium-68, which possesses a half life of 68 minutes. This is short enough for the treated fluid to reduce its activity in a reasonable time (one thousandth original activity in twelve hours) but of sufficient length to permit preparation of the isotope for incorporation in the fluid without losing too much original activity through decay. Another reason for its adoption is that considerable work has already been conducted into its chemistry for medical applications. Aside from gallium-68, rubidium-82 and fluorine-18 are two other isotopes which have suitable properties. In particular the high activity available, and shorter half life of 1.3 minutes, makes rubidium suitable for transient studies.

5.1 Isotope Extraction, Incorporation and Activity Requirements

Gallium-68 is formed as a result of the decay of a parent isotope, germanium-68. This is normally supplied bound in a porous matrix in a shielded, semi-sealed container. Extraction of the gallium from this "generator" is achieved by flushing or eluting the matrix with a suitable reagent, in this case hydrochloric acid solution. An active eluate of acidic gallium chloride solution results, from which the gallium may be extracted into a fluid- miscible solvent by using standard chelating or solvating agents. Suitable activity for a test, one to ten millicuries, is normally contained within one to five millilitres of solution. The sensitivity of the camera system for the normal detector separation of 500mm (window to window) is 17 counts per second per microcurie in the centre of the field of view (between the detectors). The time taken to accrue a full record of 1.7 million events will depend on the distribution of isotope within the total fluid system. Typical times range from 15 to 60 minutes though satisfactory results may be obtained with fewer events, with a corresponding reduction in test time.

5.2 Flow Tracing

The means of incorporating the tracer in the fluid under investigation will depend largely upon which facet of lubrication or flow behaviour is to be visualised. The options are:

a) bulk mixing with fluid,

b) injection, continuous or pulsed.

a) May be satisfactory for small volume systems where steady state conditions are of interest. Here the average activity of the fluid is made high so that a complete picture of the flow regime can be rapidly achieved.

b) Continuous injection will be of more value where a high volume complex lubrication or flow system is under investigation. The isotope will normally be injected slightly 'up stream' of the area of interest. With activity concentrated only in this area confusion is minimised. Dilution of this activated flow will then take place within the bulk so that the background level of activity produced as the fluid recycles is kept to a minimum and image quality is maintained. Pulsed injection also offers the option of being synchronised with an aspect of system operation, which may provide more easily interpreted results.

5.3 System Performance

The two most important camera performance parameters - sensitivity and resolution - were determined. Sensitivity of the camera is approximately 17 events per second per microcurie of activity in the centre of the target area, for a detector separation of 500mm. This gives information on activity versus test duration relationships for future work. Spatial resolution may be defined as the separation at which two discrete sources can just be distinguished, and so gives an indication of the fineness of features which may be seen. In the plane of the detectors (the X and Y plane in Fig. 1) it is 4.6mm. In the Z direction, perpendicular to the detectors, it depends on detector separation but at 500mm is approximately 10mm. Normally this limitation can be effectively overcome by the careful positioning of components to be investigated.

6. TEST WORK

A series of simple demonstration experiments were conducted, using gallium-68 as tracer in various hydrocarbon base fluids.

6.1 Inclined Plane Rig

Fig. 3 shows the distribution of lubricant within a sealed case containing a series of shallow inclined planes. Activated lubricant was pumped into the top of the case (arrowed) and allowed to flow to the base from where it was extracted and recirculated. Imaging was conducted through the vertical centre plane. The presence of a thin layer of oil, typically 0.5 to 0.25mm is clearly 'visible' on the surfaces. A greater thickness on the more horizontal plane, (top) is also evident. Fig. 4 shows the same rig through 25mm of steel. The oil films remain clearly delineated indicating the potential of the system for studying films in industrial objects.

6.2 Oil Wedge

Fig. 5 shows a simple wedge of activated fluid being imaged through a total of 76mm of aluminium (38mm on each side of the wedge). The length of the wedge is 150mm with fluid thickness varying from zero on the left (marked by a vertical line) to 3mm on the right. The imaging plane passes through the centre of the wedge. The histogram of counts per pixel versus film thickness yields a near linear relationship indicating that it is possible to measure film thicknesses down to 300 microns in situations of this kind. However improved film thickness resolution can be readily achieved by using higher activity levels or longer exposures.

6.3 Static Model Bearing

A model journal bearing was constructed of mild steel to provide a more realistic target for imaging. Dimensions were: height 47mm; journal diameter 40mm; total model diameter 60mm, total bearing clearance 500 microns. Imaging was conducted with the bearing's axis of 'rotation' vertical, and in the X Y plane passing through it: Fig. 6 shows the situation with the journal centralised, whilst Fig. 7 shows the journal off-centre, simulating load. In Fig.6 the distribution of activity displayed in histogram form indicates equal oil film thicknesses of 250 microns either side. The histogram of Fig. 7 shows uneven peaks. The ratio of these indicates oil film thicknesses of 125 and 375 microns.

7. FUTURE PROSPECTS

A full test programme has been initiated to investigate potential areas of application. These include flow splitting and oil stagnation and hiding, foaming, mixing and separation of bulk fluids, and thin film visualisation under realistic conditions. Such aspects have their applications in many areas of lubrication, cooling and fuel supply. The simplified examples studied to date have allowed valuable experience to be generated in camera operation which will be applied to more realistic situations as the work progresses. Continuing improvements in data analysis techniques and electronics are also expected to improve and expand the industrial application of PET.

ACKNOWLEDGEMENTS

This project was funded by a Science and Engineering Council cooperative grant, and by the joint industrial partners Castrol Limited and Rolls Royce Limited. The apparatus was built at the Rutherford Appleton Laboratories and experimental work was conducted in the Department of Physics Radiation Centre, University of Birmingham.

Aspects of this equipment and techniques used are covered by patent applications.

REFERENCES

1. STEWART P.A.E. Aero Engine Applications of Cold Neutron Fluoroscopy at Rolls Royce Limited. Neutron Radiography Proc first World Conf. San Diego, Ca., U.S.A., 1981. Eds. J.P. Barton & P von Der Hardt; D. Reidl, Dordrecht, Holland 1983 (P625).

2. STEWART P.A.E., HERITAGE. J. Cold Neutron Fluoroscopy of Operating Automotive Engines. Ibid (P635).

3. BATEMAN J.E. et al. The Rutherford Appleton Laboratory's Mark I Multiwire Proportional Counter Positron Camera. Nuclear Insts. and Meths. (1984) 225, 209-231.

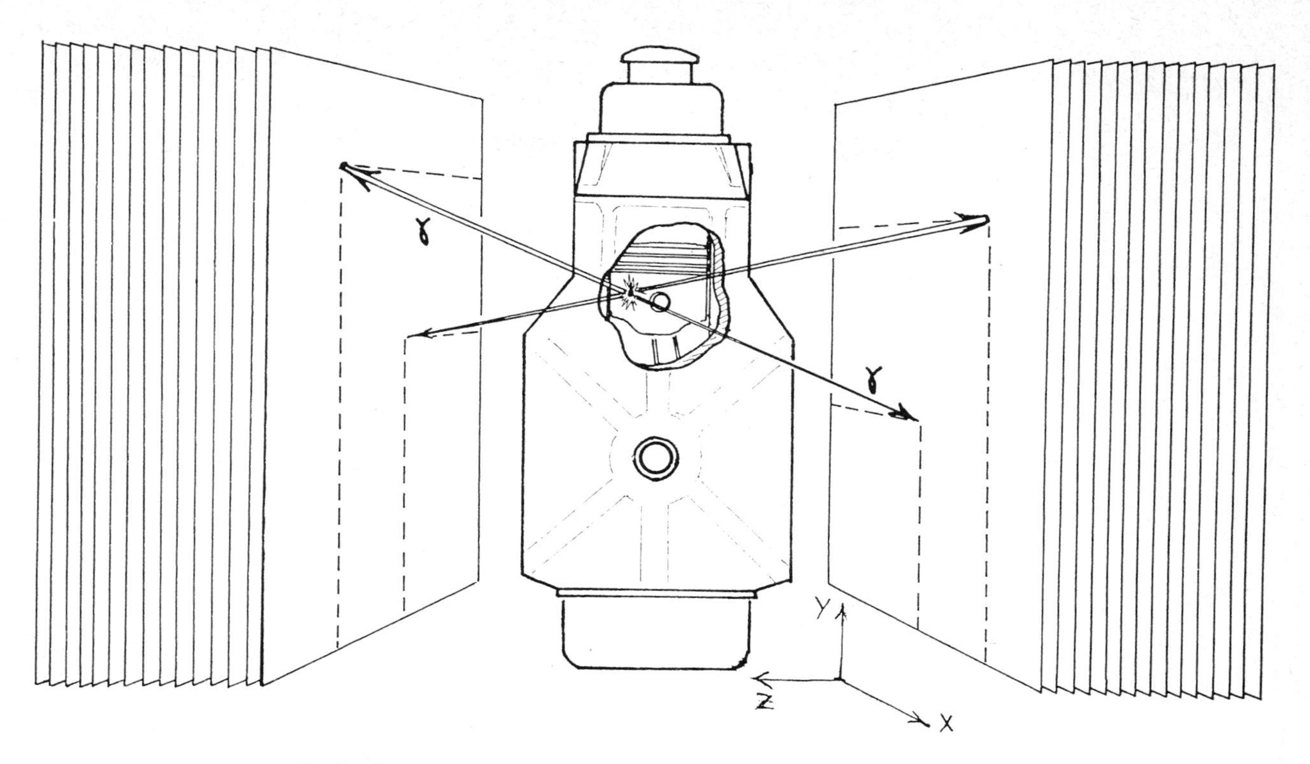

Fig 1 Principle of operation of Positron Emission Tomography (PET)

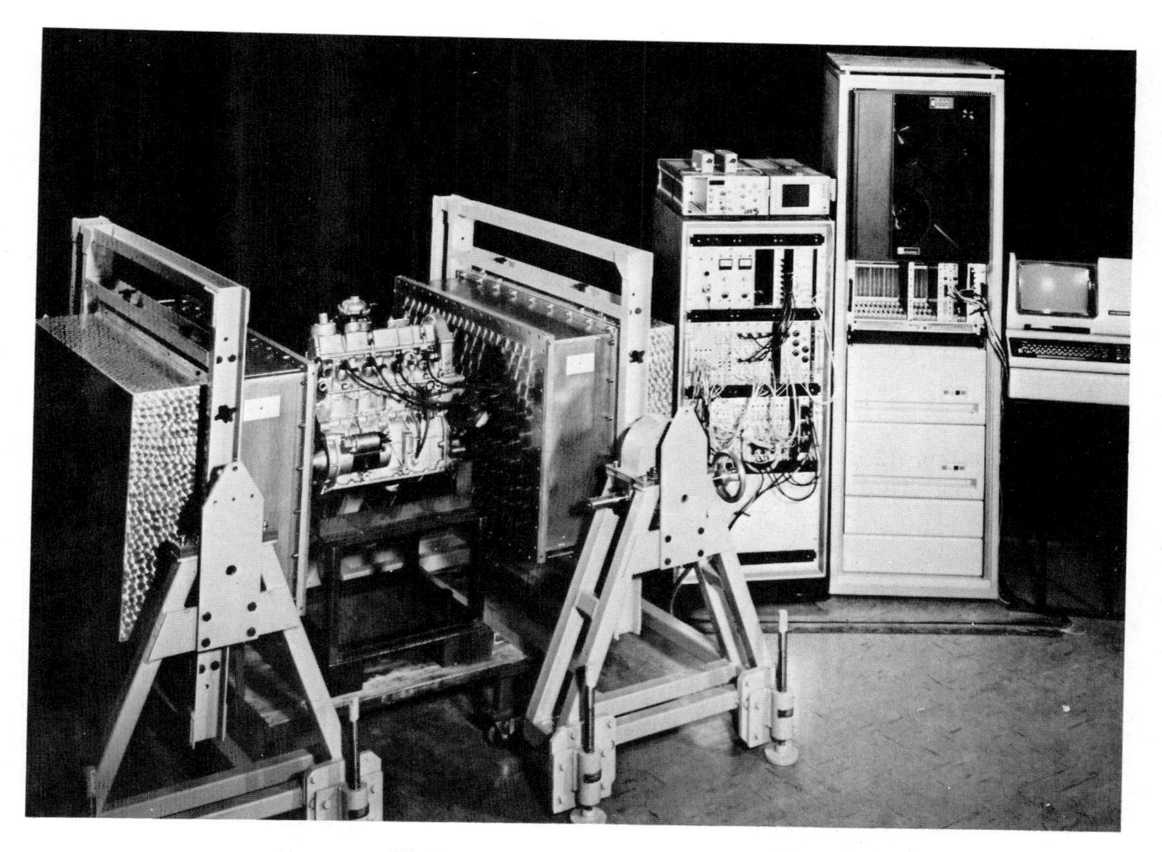

Fig 2 The industrial PET camera system

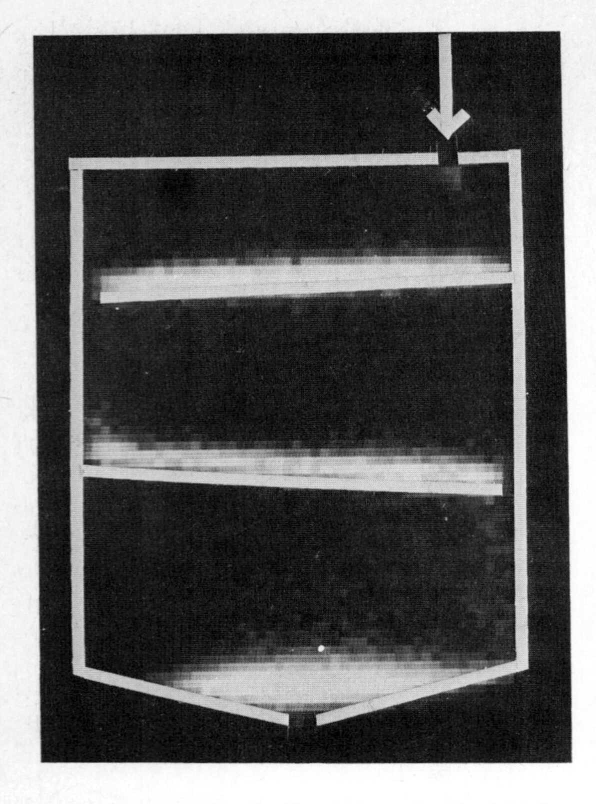

Fig 3 Flow tracing: inclined plane rig. No shielding

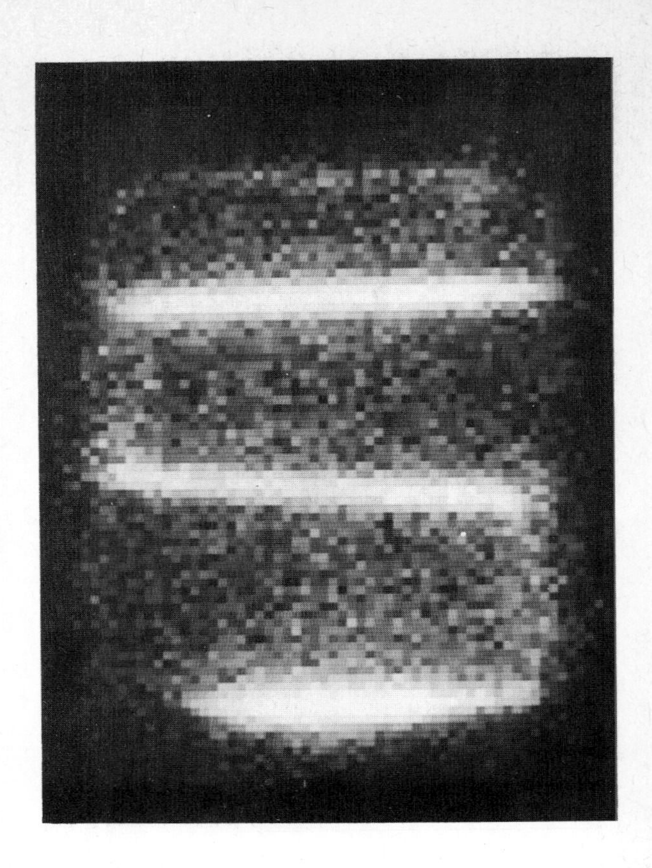

Fig 4 Flow tracing: inclined plane rig. 25 mm steel shielding between target and detectors

Fig 5 Oil wedge. 76 mm aluminium between target and detectors

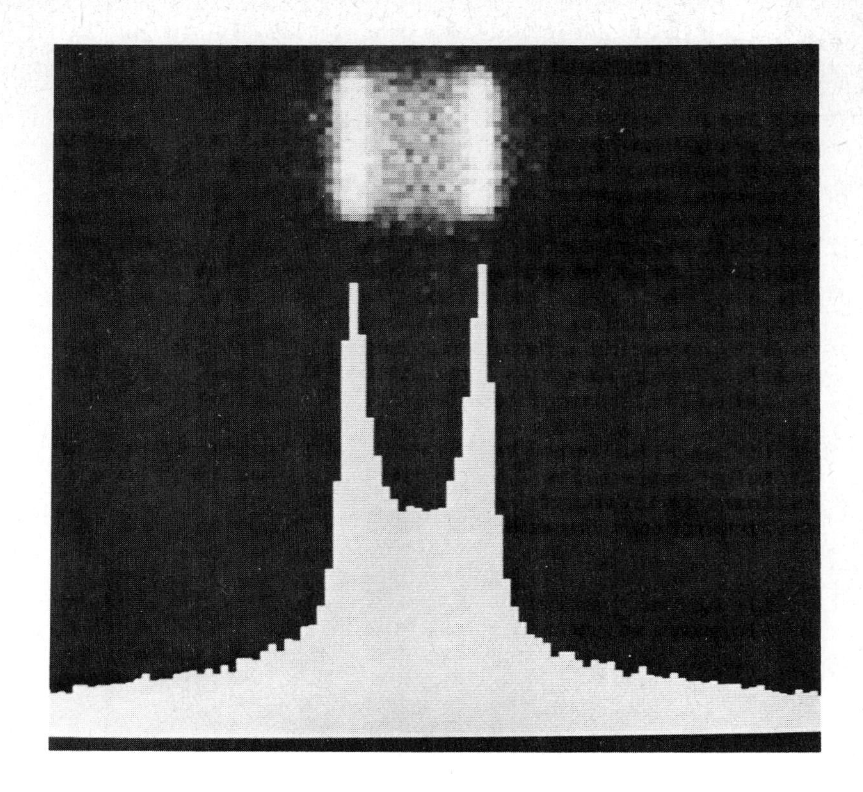

Fig 6 Model bearing – centralized. Equal film thickness indicated by activity histogram. (Total clearance 500 μm)

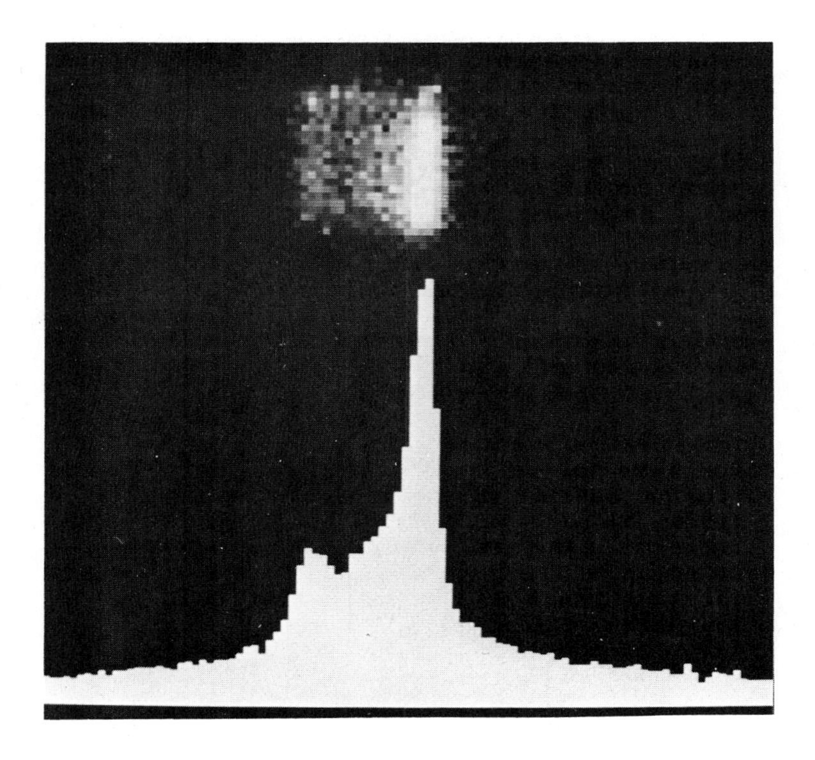

Fig 7 Model bearing – off centre. Indicated film thickness ratio 3:1 (375 and 125 μm)

The effect on friction of lubricants containing zinc dithiophosphate and organo-molybdenum compound

K KUBO, BSc, MSc, M KIBUKAWA, BSc and Y SHIMAKAWA, MSc, PhD
Shell Sekiyu KK, Kanagawa-ken, Japan

SYNOPSIS The frictional behaviour of mineral oil containing several primary or secondary zinc dithiophosphates (ZDTPs) individually and in combination with a friction reducing organo-molybdenum compound (OMC) has been investigated in the Amsler friction and wear tester and the worn surfaces of the pins have been examined by electron probe micro-analysis. The friction coefficient of ZDTP-containing oil is sensitive to ZDTP type and is not directly related to the concentration of the surface film. The addition of OMC to reactive ZDTP solutions is effective in decreasing the friction coefficient but it remains sensitive to ZDTP type.

1 INTRODUCTION

ZDTPs (zinc dialkyl dithiophosphates) have been successfully used as antiwear and antioxidation additives in lubricants for a long time and their mechanisms of action have been studied by many investigators. Although many papers have discussed the various functions of ZDTPs such as decomposition (1-3) and surface film formation (4-8), frictional changes on the addition of ZDTPs (9) etc., the mechanisms have not been fully clarified. In addition, friction modifiers (FMs) have become of increasing interest in the last few years in order to minimize boundary friction, and hence ZDTPs have been used with the addition of friction modifiers. Even though ZDTPs have been used with such friction modifiers and the effectiveness of various combinations noted, not many studies have been reported.

It has been postulated that the mechanisms involved in reducing friction and wear for ZDTP/FM combinations are complicated, nevertheless it is important to study the behaviour of ZDTPs with and without friction reducing additives. Accordingly, the authors have investigated the frictional behaviour of ZDTPs with and without an organic molybdenum compound (OMC) under boundary lubrication conditions in an Amsler friction and wear test machine. The worn surfaces have been analysed in an electron probe microanalyser (EPMA) and the composition of the surface films determined. The results of the investigation are discussed and explanations of the action of ZDTP and OMC are presented.

2 EXPERIMENTAL

An Amsler friction and wear test machine, which is shown schematically in Fig. 1, was employed to investigate the friction characteristics of various lubricating oils.

The stationary test piece which has a 2 x 8 mm contact area and is 15mm in length is loaded against a rotating disc of 50mm diameter and 13mm width. They are both of hardened S45C carbon steel with HV 214-216 and HV 560-590 hardness respectively. Prior to the friction test, the stationary test piece mounted on the upper shaft in a holder was rubbed against a special run-in disc for 20 minutes under a load of 600N to give a conforming surface of 2 x 8 mm^2. A paraffinic mineral oil was used to lubricate the contact area. After obtaining a conforming surface on the stationary test piece, a test disc with surface roughness controlled within the range 0.13 to 0.17 µm Ra was mounted on the lower shaft. 1000ml of test oil controlled at 70°C was circulated via a jet into the inlet of the contact at a feed rate of 290ml/min. The disc was rotated with no load for 5min. followed by step loading at 60N/5sec. up to 600N. The rotating speed of the lower disc was kept at 200rpm (0.52m/s) sliding speed throughout the friction tests. The friction coefficients of the test oils were calculated from the transmitted friction torque. The friction torque measuring device, mounted between the electric motor and the shaft carrying the lower test specimen, was a gear dynamometer with a pendulum which swung out of the vertical balanced position when power was transmitted. The sine of the angle of inclination was a measure of the torque.

The composition of the ZDTP and MoDTP compounds used in this work is shown in Table 1. The metal contents were obtained using atomic absorption spectroscopy. The experimental zinc contents are in good agreement with the values calculated on the basis of the ZDTP structures indicating high sample purity. Decomposition temperatures measured using differential thermal analysis (DTA) are also presented in Table 1.

The oils used in the test programme were blends of paraffinic mineral oil of viscosity 56 cSt and 8.0 cSt at 40°C and 100°C respectively and ZDTPs and an organic molybdenum compound. The Zn and Mo concentrations were usually 0.1 per cent wt. and 400ppm wt. respectively. Some other concentrations of ZDTPs and OMC (see Table 4 and 5) were used to investigate the influence of concentration.

Electron probe microanalysis (EPMA) was applied to characterize the rubbed surfaces of the stationary test pieces in some experiments. The concentrations of the elements reported in Tables 2 - 5, are the mean values of counts obtained at 70 spots of 80 μm ϕ dotted along the centre line of the rubbed surface. In general the standard deviation was about 10 per cent of the mean value. No corrections for surface roughness, composition etc. were made.

The authors have recognized that, in practice, the rubbing conditions such as combination of materials, contact load, sliding speed, temperature etc. can influence the frictional results. However, at this initial stage of our investigations of ZDTPs with other additives, the test conditions were fixed.

3 RESULTS AND DISCUSSION

3.1 Frictional characteristics of ZDTP containing oils

The first experiments to be described were carried out in the Amsler test machine using different ZDTPs in base oil at 0.1 percent wt. zinc concentration with and without OMC at 400 ppm molybdenum. The friction results are plotted in Fig. 2 (a) and (b). The frictional characteristics were sensitive to the type of ZDTP used in the tests and there was no relation between primary or secondary ZDTP and friction coefficient. The friction coefficients obtained with pC_8, pC_{12} and sC_6 were relatively high whereas those of pC_5 and sC_8 were low. The friction obtained with pC_8 and pC_{12} also fluctuated periodically, the period slowly increasing with time to a steady value of approx. 20 minutes. Base oil containing OMC and the oil with sC_6 exhibited relatively high friction. Base oil on its own gave rapid fluctuations in friction (as illustrated in Fig. 2a) and this was associated with high wear.

It is generally considered that the decomposition temperatures of secondary type ZDTPs are lower than those of primary type ZDTPs. However, no relationship between the decomposition temperature assessed by differential thermal analysis (DTA) and the corresponding frictional characteristics was obtained. For example, the decomposition temperature of sC_6 was 175°C by DTA (see Table 1) measured in a nitrogen atmosphere and was only marginally lower than the 178°C decomposition temperature of sC_8 although there was a remarkable difference in friction coefficient between these two ZDTPs. Again, the friction coefficient of pC_8 was lower than that of sC_6 which was a steady high value whereas the decomposition temperature of pC_8, 213°C, is much higher than that of sC_6, 175°C.

The results of surface analysis performed after 120 minutes rubbing are listed in Table 2. Figures in parenthesis are relative intensities normalized to phosphorus. There does not appear to be a relationship between Zn, P or S intensities and the DTA decomposition temperatures. For example, a comparison between sC_6 and sC_8 whose decomposition temperatures are very similar (175°C and 178°C respectively), give quite different elemental intensities and ratios.

In these experiments, it is also clear that there is no relation between friction coefficient and reactivity or film forming tendency of ZDTPs as assessed by EPMA. For instance, when we consider the change of friction and element intensities of pC_8, the higher the friction coefficient, the higher the element intensities in the film {compare pC_8 (↑) and pC_8 (↓) in Table 2}. It should be noted that the element ratios for Zn and P did not change appreciably with change in intensity, however the sulphur and oxygen ratios did vary greatly.

Regarding the different ZDTP types, primary ZDTPs gave higher sulphur ratios than secondary ZDTPs. The total amount of reacted film doesn't correlate with the effectiveness of ZDTPs in reducing friction.
In general, it seemed that the higher friction coefficients were accompanied by higher element intensities which suggests that thicker films were formed although there was an exceptional result with pC_5. It has also been reported by others that increased friction was associated with the addition of ZDTP to base oil (9).

Assuming this trend is correct, the friction coefficient should correlate with the total amounts of reacted layer and thin surface layers should give lower friction.

3.2 Frictional characteristics of oils containing ZDTP and OMC

In the second series of experiments, test oils containing ZDTP at 0.1 per cent wt. Zn and OMC at 400ppm Mo were examined in the same friction test. Results of the friction tests and the surface analyses are given in Fig. 3 and Table 3.

The fluctuating friction coefficients exhibited when pC_8 and pC_{12} were used without OMC disappeared when evaluated with the friction modifier. OMC showed rather high friction coefficient when used alone (Figure 2), however, there was synergistic effect on friction with pC_8 and sC_6 (Figure 3). With the other ZDTPs, which gave lower friction when used alone, the addition of OMC tended to increase the friction to a level somewhere between that of the ZDTPs alone and OMC alone. Again, even with OMC present, the friction coefficients obtained were sensitive to ZDTP type. In the case of pC_5 the friction ran very high and the experiment was stopped due to scuffing. As can be seen by comparing the results in Tables 2 and 3, the element intensities of the films were decreased by the addition of OMC except in the case of sC_8 where they were increased. The elemental ratios of the FM and non-FM oils were similar for pC_8, pC_{12} and sC_6 experiments and thus it is considered that the structure or chemical composition of the reacted surface films in the FM version are probably similar to those formed in the non-FM experiments. In the case of the FM version of sC_8, the intensities were increased by the addition of OMC and the element ratios were changed. This might be the result of a different film structure.

Friction results of experiments with pC_8 and sC_6 oils were so remarkably decreased by the addition of OMC that both FM versions of these two ZDTPs were investigated with various

OMC concentrations. The results are shown in Fig. 4 with surface analysis shown in Table 4. As the OMC concentration increased, the friction coefficient steadily decreased in both cases. From these results, it is clear that the addition of OMC is effective when it is added to ZDTP solutions which gave high friction coefficients in the first place.

3.3 The influence of ZDTP concentration on the frictional behaviour of FM oils

In this final series of experiments, the ZDTP concentration in the FM oils was altered while keeping the OMC concentration fixed at 400ppm wt of Mo. Results for pC_8 and sC_6 are shown in Fig. 5 and Table 5. In the range of 0.025 to 0.1 per cent wt. of Zn, pC_8 plus OMC was effective in maintaining low friction whereas at 0.2 per cent wt. Zn, the friction was high. For sC_6 rather higher friction levels were obtained at low concentrations of Zn and low friction was only obtained at Zn concentrations of 0.1 and 0.2 per cent wt.

Surface analysis revealed that the relatively low intensities of Zn, P and Mo in the case of 0.2 per cent pC_8 and 0.025 per cent sC_6 were associated with rather high friction coefficients. pC_8 was more effective than sC_6. The intensity distributions measured across the width of the rubbed surfaces, as shown in Fig. 6, suggest that Zn, P and Mo have similar distributions in the case of pC_8, (sC_6 also gave similar results) whereas in the case of sC_8, Zn and P have similar distributions but that of Mo is different. This coexistence of Mo with the elements originating from the ZDTP in the case of pC_8 and sC_6 suggests that the reactions of ZDTP and of OMC occur at the same spots in the contact zone and lead to films giving low friction. On the other hand, where Mo is not associated with Zn and P on the surface, high friction is observed.

3.4 The influence of concentration of ZDTPs and OMC

The intensities of the elements Mo and Zn on the rubbed surfaces and the friction levels observed are summarized in Fig. 7.

From the results for sC_6 summarized here, it is apparent that for a Zn concentration of 0.1 per cent wt., an increase of OMC concentration results in a decrease of reacted Zn film. For pC_8 this trend is less clear since for Mo at 100pm the Zn concentration falls below that at 400ppm Mo concentration. Considering the double circles which show relatively low friction, it appears necessary to add at least 400ppm Mo as OMC to give Mo intensities on the rubbed surfaces of more than 100 counts. From Fig. 7 it can been seen that Zn intensities on the rubbed surfaces do not characterize the level of friction. It is probable that low friction relates more to Mo concentration than Zn concentration on the rubbed surfaces but that it is essential to have some zinc in the surface films.

The final experiment in the Amsler test machine was carried out to assess the characteristics of reacted films formed by oils containing sC_6 and OMC. The results in Fig. 8 (a), (b) indicate the degree of friction change when the additive in the test oil was replaced by either OMC or sC_6. Replacing sC_6 by OMC gave a very low friction coefficient, however replacing OMC by sC_6 did not. These results indicate that the original reacted film of OMC was not durable and requires replenishing to maintain decreased friction whereas the original sC_6 derived film is durable.

In addition to these experiments, the results shown in Fig. 8 (c) were obtained when sC_6-OMC oil was replaced by base oil. There was an immediate and rapid decrease in the intensities of each element and a sharp rise in friction. The intensities 2 minutes after replacement {(2) in diagram} were nearly the same as those after 10 minutes {(3) in diagram}. Even though the Mo intensity was not low, a low friction coefficient could not be maintained. Zn and P element intensities after replacement were about a half of those before.

This suggests that the reacted films of sC_6 and OMC do not exist separately in a microscopic sense and that the momentary film or layer at the outer surface plays a role in decreasing the friction.

On the other hand, as seen in Fig. 8 (a), OMC replacing sC_6 decreased the friction successfully in spite of its relatively low Mo intensity which was less than 100 counts and from the previous experiments was considered to be too low to be effective. This suggests that the thin outer surface layer abounds in Mo, whereas the Mo counts for the whole film are relatively small. However, when the OMC has been added to sC_6 solution instead of keeping replaced, the Mo count was over 100 as seen in the sC_6-OMC version in Fig. 8 (c) and the friction was low.

Fig. 8 also shows a conceptual scheme which illustrates the case of OMC (b-1), OMC plus sC_6 (c-1) and OMC replacing sC_6 (a-2). Case (c-1) is considered to be reasonable as a result of competitive reaction of OMC and sC_6 and each of the intensities are less than the values obtained in oils containing only one additive. Unexpectedly low Mo intensity shown in (a-2) could be considered to result from a lower probability of OMC to react with steel surfaces or by rapid removal from Zn rich surface films. Considering its remarkably small Mo intensity, a low friction coefficient was observed in case (a-2), whereas the ZDTP-OMC version needed more than 100 counts of Mo to give the same level of friction coefficient. This suggests a probability of low friction at the contact surface layer which is brought about by low shear stress at the interface of the reacted Zn rich layer and thin Mo rich film at the surfaces.

4 CONCLUSIONS

The frictional characteristics of oils which contained ZDTPs with and without OMC as friction modifier were studied in the Amsler friction and wear test machine and the results suggest that

a) The measured friction is sensitive to ZDTP type with or without the addition of friction modifier.

b) The friction coefficients of oils which contain ZDTP as additive do not directly relate to the concentration of the surface films formed and some ZDTPs show relatively high friction coefficient with rich surface reacted films.

c) The addition of OMC to reactive ZDTP solutions is effective in decreasing the friction and indicates that an OMC concentration of more than 400ppm Mo should be used.

d) There is continuous reaction of OMC which means sufficient replenishment is required to decrease the friction whereas the ZDTP derived film is relatively durable.

e) Competitive reaction exists between ZDTP and OMC, and there is an obvious decrease in reacted Zn layer caused by OMC concentration.

f) It is possible that the friction is characterized by the Mo rich layer at the surface, and that the lower friction coefficient is due to the easily shearable structure of a reacted Zn rich layer surfaced by a thin Mo rich layer.

REFERENCES

(1) GALLOPOULOS, N.E. Thermal decomposition of metal dialkyldithiophosphate oil blends. ASLE Transactions, 1964, 7, 55-63.

(2) FENG, I.M. Pyrolysis of zinc dialkyl phosphorodithioate and boundary lubrication. Wear, 1960, 3, 309-311.

(3) JONES, R.B. and COY, R.C. The thermal degradation and EP performance of zinc-dialkyldithiophosphate additives in white oil. ASLE Transactions, 1981, 24, 77-90.

(4) FENG, I.M., PERILSTEIN, W.L. and ADAMS, M.R. Solid film deposition and non-sacrificial boundary lubrication. ASLE Transactions, 1963, 6, 60-66.

(5) ROWE, C.N. and DICKERT, J.J. The relation of antiwear function to thermal stability and structure for metal O,O-Dialkylphosphorodithioates. ASLE Transactionss, 1967, 10, 85-90.

(6) COY, R.C. and Quinn, T.F.J. The use of physical methods of analysis to identify surface layers formed by organosulfur compounds in wear test. ASLE Transactions, 1975, 18, 163-174.

(7) BARCROFT, F.T., BIRD, R.J., HUTTON, J.F. and PARK, D. The mechanism of action of zinc thiophosphates as extreme pressure agents. Wear, 1982, 77, 355-384.

(8) BIRD, R.J. and GALVIN, G.D. The application of photoelectron spectroscopy to the study of EP films on lubricated surfaces. Wear, 1976, 37, 143-167.

(9) RUDIGER, H. The influence of boundary layers on friction. Wear, 1979, 56, 147-154.

(10) HAVILAND, M.L. and GOODWIN, M.C. Fuel economy improvements with friction modified engine oils in environmental protection agency and road test. SAE 790945.

(11) BRAITHWAITE, E.R. and GREENE, A.B. A critical analysis of the performance of molybdenum compounds in motor vehicles. Wear, 1978, 46, 405-432.

(12) HAMAGUCHI, H., MAEDA, Y., and MAEDA, T. Fuel efficient motor oil for Japanese passenger cars. SAE 810316.

(13) RETZLOFF, J.B., DAIVIS, H.E., and GLUCKSTEIN, M.E. Fuel economy benefits from modified crankcase lubricants. ASLE Transactions, 1979, 35, 568-576.

Table 1 Composition of zinc-dialkyldithiophosphates (ZDTPs) and molybdenum dialkyldithiophosphates used in this work

Code	Alcohol type	Concentration, %w				Initial decomposition temperature, °C [1)]
		Zn*	P	S	Mo	
pC_8	2-ethylhexyl	8.33 (8.46)	7.82	15.6	-	213
pC_{12}	n-dodecyl	6.58 (6.56)	6.18	12.5	-	-
pC_5	n-pentyl	10.70 (10.80)	10.10	20.3	-	194
sC_6	4-methyl 2-pentyl	9.69 (9.90)	9.58	19.0	-	175
sC_8	2-octyl	8.55 (8.46)	7.88	15.5	-	178
OMC [2)]	-	-	6.20	13.60	7.9	175

* Figures in parenthesis are calculated concentrations.

1) Differential thermal analysis conditions

Standard sample : Al_2O_3
Sample size : 8mg Approx.
Heating rate : 5°C/min
Temp. range : 25 - 450°C
Atmosphere : N_2

2) Main components of OMC

50%, Sulfurized oxy molybdenum dithiophosphate

$$\left[O{=}Mo\langle^{S}_{S}\rangle Mo{=}O \right]\left[-S-\overset{\overset{S}{\|}}{P}(OR)_2 \right]_2 \quad (R : 2\ Ethyl\ hexyl)$$

8.4%, Mono, Di-acid dithiophosphate ester

$$(RO)_m-\overset{\overset{S}{\|}}{P}-(OH)_n \quad (n : 1\ or\ 2,\ m + n = 3)$$

19.1%, Unknown

22.5%, Mineral oil

Table 2 EPMA results of rubbed surfaces.
Lubricant Zn concentration 0.1%w

(400ppm Mo for last test)

Additive component	Zn Lα	S Kα	P Kα	Mo Lα	O Kα
pC_8 (↓)[1]	540 (2.3)[3]	1500 (6.5)	230 (1)	<1 (0)	130 (0.6)
pC_8 (↑)[2]	1700 (2.5)	2600 (3.8)	680 (1)	<1 (0)	160 (0.2)
pC_{12}	1800 (2.3)	5500 (7.1)	770 (1)	<1 (0)	120 (0.2)
pC_5	1800 (3.7)	2000 (4.1)	490 (1)	<1 (0)	120 (0.2)
sC_6	2800 (2.1)	1300 (1.0)	1300 (1)	<1 (0)	200 (0.2)
sC_8	490 (3.5)	390 (2.8)	140 (1)	<1 (0)	110 (0.8)
OMC	<1 (0)	430 (0.5)	820 (1)	320 (0.4)	180 (0.2)

1) pC_8 (↓) measured at low μ point.

2) pC_8 (↑) measured at high μ point.

3) Figures in parenthesis normalised to phosphorus

Table 3 EPMA results of rubbed surfaces.
Lubricant Zn concentration 0.1%w,
Mo concentration 400ppm

Additive component	Zn Lα	S Kα	P Kα	Mo Lα	O Kα
pC_8 + OMC	1400 (2.1)[1]	2900 (4.3)	670 (1)	180 (0.3)	110 (0.2)
pC_{12} + OMC	620 (1.9)	3000 (9.4)	320 (1)	120 (0.4)	100 (0.3)
sC_6 + OMC	1200 (1.8)	1200 (1.8)	680 (1)	120 (0.2)	130 (0.2)
sC_8 + OMC	770 (1.9)	560 (1.4)	410 (1)	81 (0.2)	120 (0.3)

1) Figures in parenthesis normalised to phosphorus

Table 4 EPMA results of rubbed surfaces. Lubricant Zn concentration 0.1%w, Mo concentration varied

Additive component	Zn $L\alpha$	S $K\alpha$	P $K\alpha$	Mo $L\alpha$	O $K\alpha$
pC_8 + OMC					
Zn + Mo, 100ppm	1000 (2.9)	1200 (3.5)	340 (1)	16 (0.05)	220 (0.6)
Zn + Mo, 400ppm	1400 (2.1)	2900 (4.3)	670 (1)	180 (0.3)	120 (0.2)
Zn + Mo, 1600ppm	200 (1.3)	490 (3.3)	150 (1)	140 (0.9)	97 (0.7)
sC_6 + OMC					
Zn + Mo, 100ppm	2600 (2.2)	850 (0.7)	1200 (1)	16 (0.01)	190 (0.2)
Zn + Mo, 400ppm	1100 (2.3)	1200 (2.3)	480 (1)	120 (0.3)	130 (0.3)
Zn + Mo, 1600ppm	340 (1.2)	670 (2.4)	280 (1)	160 (0.6)	120 (0.4)

1) Figures in parenthesis normalised to phosphorus

Table 5 EPMA results of rubbed surfaces. Lubricant Mo concentration 400ppmw, Zn concentration varied

Additive component	Zn $L\alpha$	S $K\alpha$	P $K\alpha$	Mo $L\alpha$	O $K\alpha$
pC_8 + OMC					
Zn, 0.025%w + Mo	810 (1.5)[1)]	1600 (3.0)	530 (1)	200 (0.4)	150 (0.3)
Zn, 0.10%w + Mo	1400 (2.1)	2900 (4.4)	670 (1)	180 (0.3)	120 (0.2)
Zn, 0.20%w + Mo	390 (2.4)	1300 (8.1)	160 (1)	67 (0.4)	110 (0.7)
sC_6 + OMC					
Zn, 0.025%w + Mo	330 (0.8)	420 (1.1)	400 (1)	59 (0.1)	130 (0.3)
Zn, 0.10%w + Mo	1200 (1.8)	1200 (1.8)	680 (1)	120 (0.2)	130 (0.2)
Zn, 0.20%w + Mo	1900 (2.2)	1200 (1.4)	860 (1)	110 (0.1)	140 (0.2)

1) Figures in parenthesis normalised to phosphorus

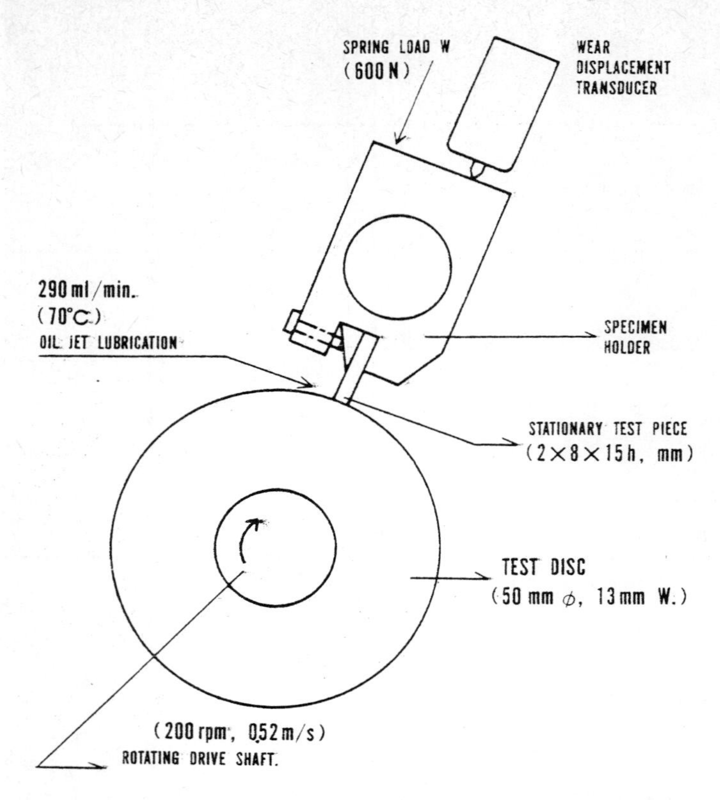

Fig 1 Configuration of Amsler friction and wear tester

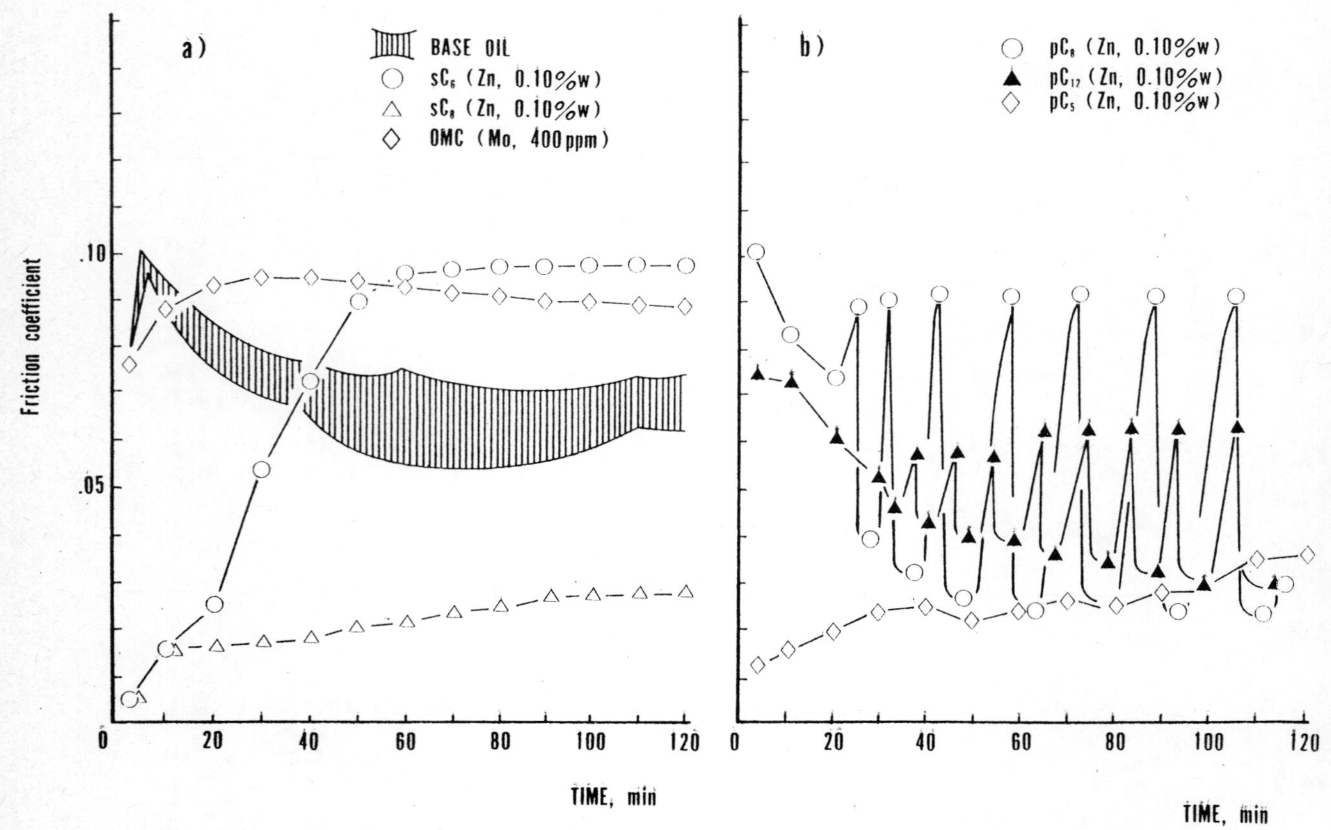

Fig 2 Friction coefficients of base oil and blends containing ZDTP or OMC

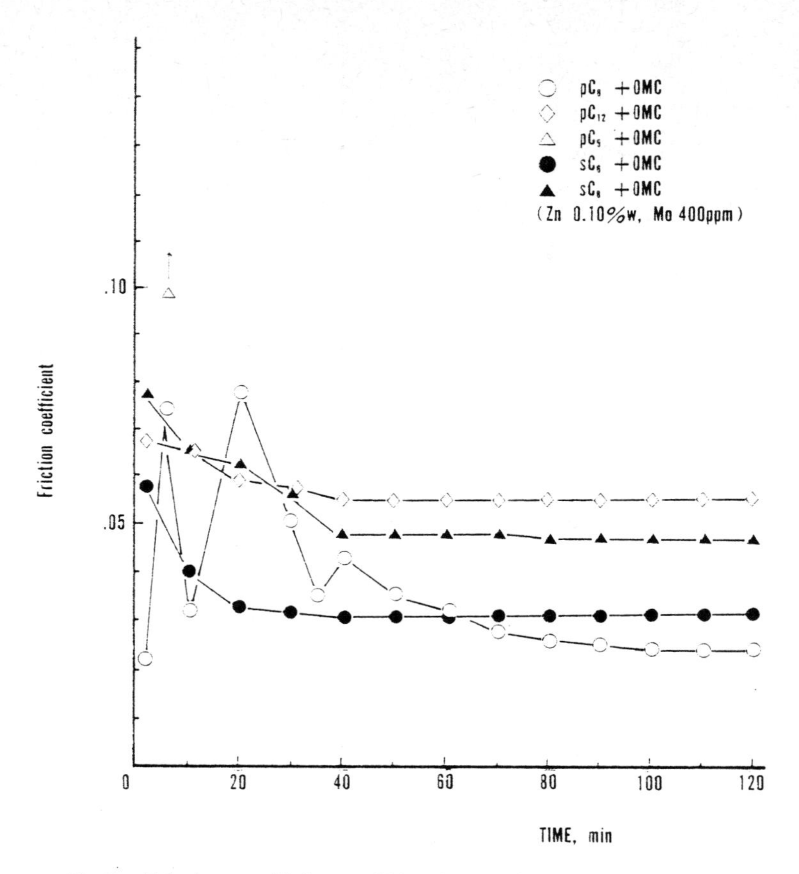

Fig 3 Friction coefficients of blends containing ZDTP and OMC

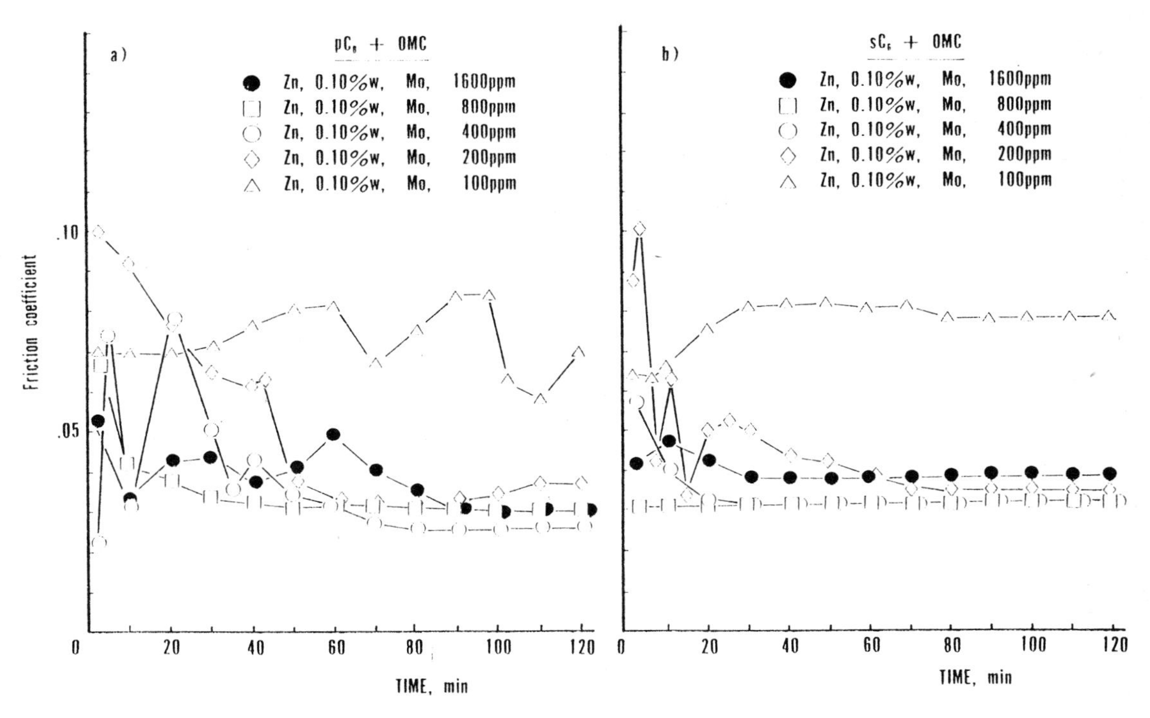

Fig 4 Friction coefficients of ZDTPs with different OMC concentrations

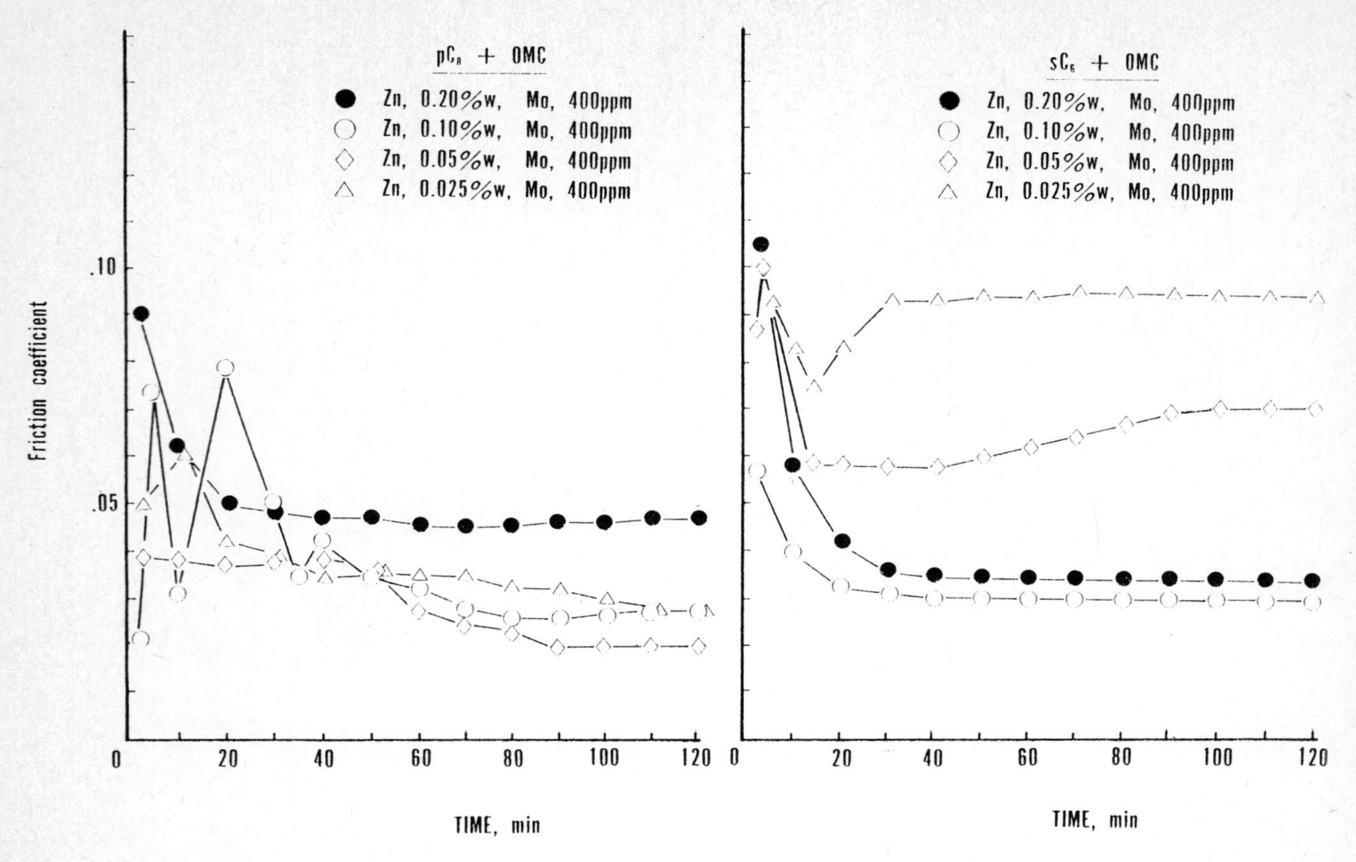

Fig 5 Friction coefficients of OMC with different ZDTP concentrations

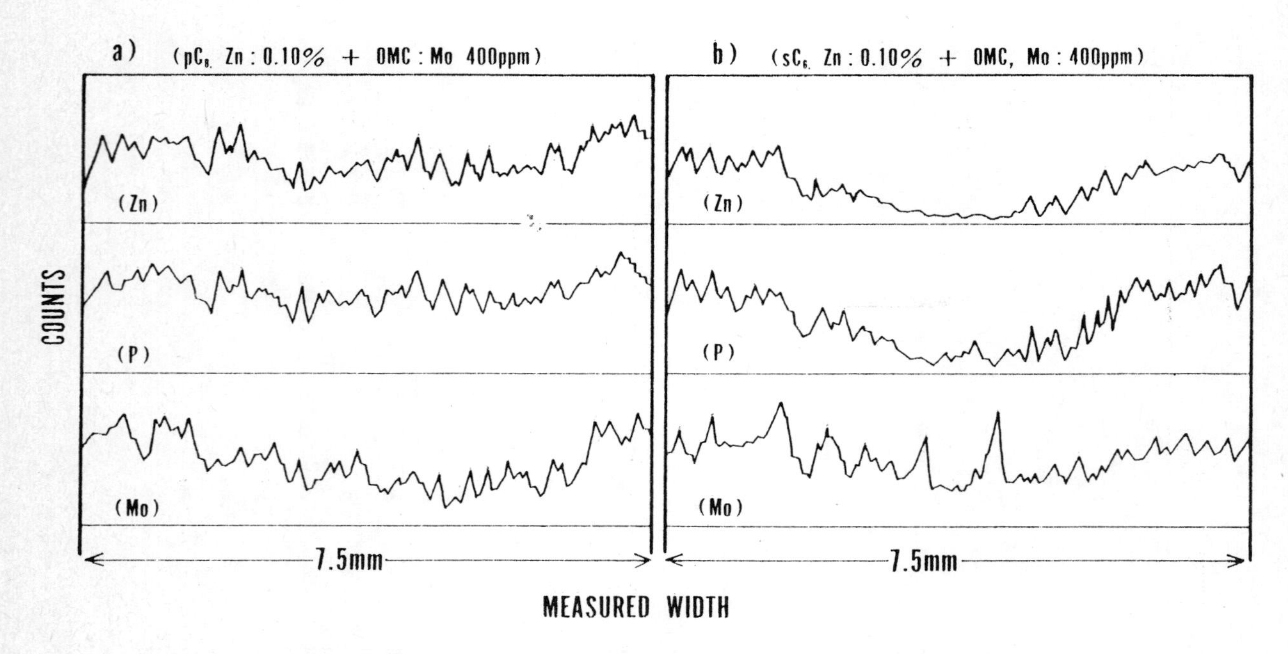

Fig 6 Element distributions along the scar width

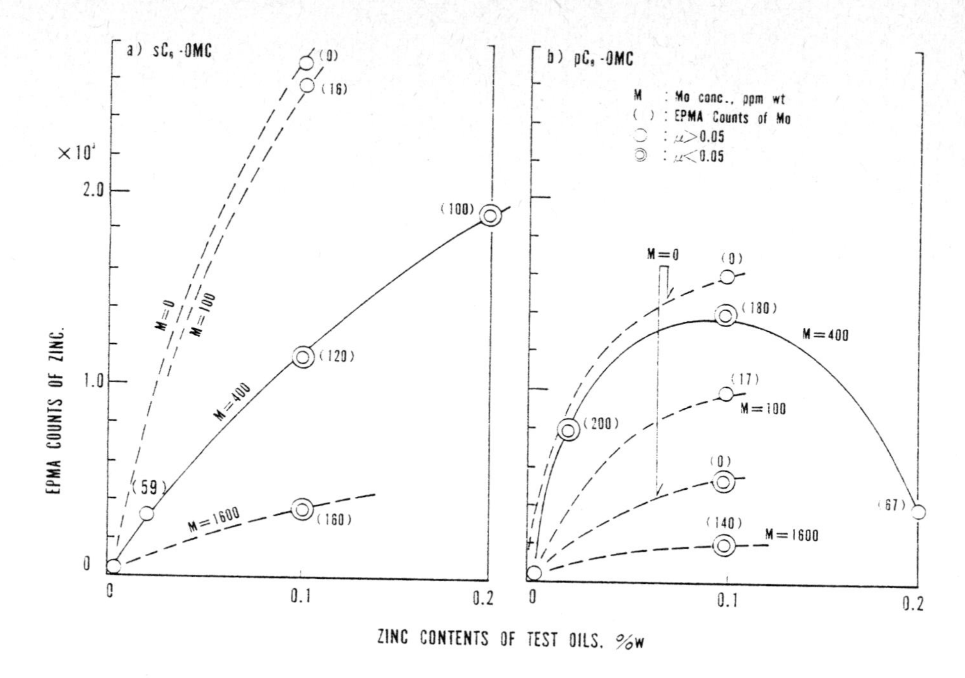

Fig 7 Relationship between additive concentration, reacted element intensities and friction coefficients

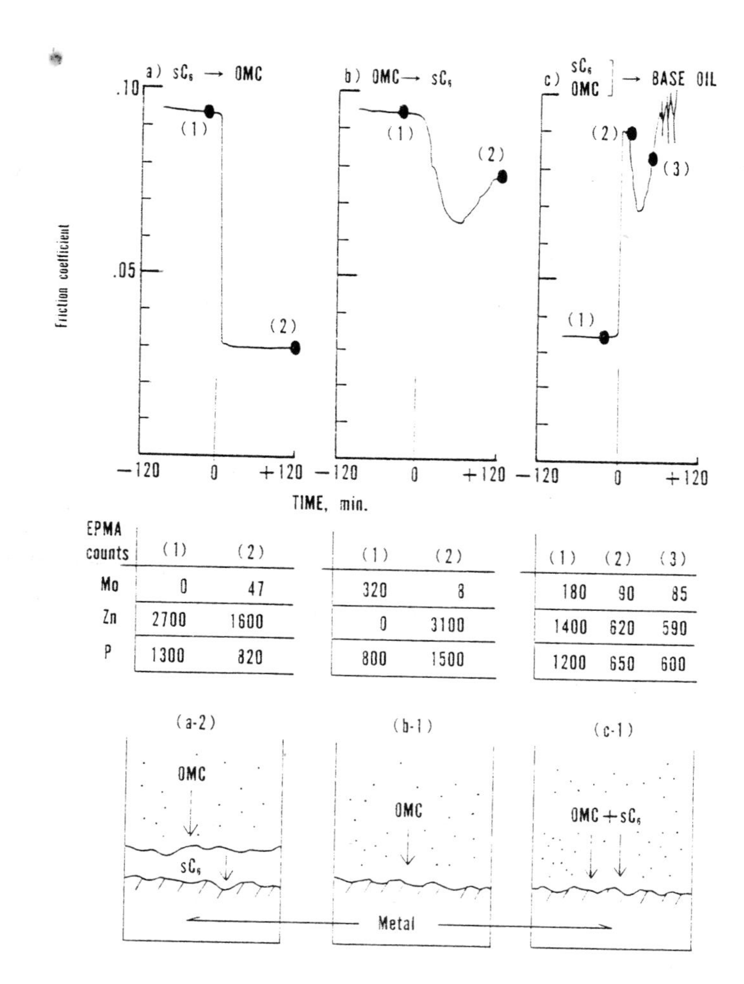

EPMA counts	a) (1)	a) (2)	b) (1)	b) (2)	c) (1)	c) (2)	c) (3)
Mo	0	47	320	8	180	90	85
Zn	2700	1600	0	3100	1400	620	590
P	1300	820	800	1500	1200	650	600

Fig 8 Relationship between friction coefficient and test oil switching procedure and conceptual illustration of reactions under boundary lubrication

Tribological behaviour of alpha silicon carbide engine components

J DERBY, BA, MS, J MacBETH, BS, MS and S SESHADRI, BE, MS, PhD
SOHIO Engineered Materials Company, Niagara Falls, New York, USA

ABSTRACT

An investigation was conducted on the wear and friction behaviour of sintered alpha silicon carbide for use in engine component applications. Laboratory and engine component tests were conducted and the effects of variables such as load and surface roughness were examined for dry and lubricated conditions. Alpha silicon carbide was found to produce lower friction and wear than standard metal valve train components. Reduced surface roughness produced significant decreases in friction for lubricated sliding. Test results are evaluated in terms of fundamental mechanisms and operating conditions.

INTRODUCTION

Engine manufacturers are continually searching for components which are economical and light weight, as well as wear and high temperature resistant to provide improved fuel economy and engine performance. Ceramics possess many of these favourable properties, and are being evaluated for engine component applications including valve lifters, seats, and guides as well as pads, push-rod tips, cylinder liners, and piston parts.

An assessment of wear and friction properties of materials used for engine components is important, yet difficult due to the diverse conditions under which they operate. Wear and friction behaviour is influenced by many variables such as: (a) the nature of material surfaces in contact including the mechanical response, surface conditions, and contact configuration, (b) lubricant properties and (c) the environment which can affect surface events such as chemisorption, compound formation, segregation or reconstruction of the atomic layers.[1] However, effects of some individual variables or isolated mechanisms on wear and friction behaviour can be studied through careful experimental design.

The particular theory of wear subscribed to such as adhesive, abrasive, or brittle fracture can influence perceptions of tribological behaviour. The adhesive and abrasive theories are well established and have been applied to metals extensively. Both theories distinguish the real area of contact, assumed to be the ratio of load to hardness, from the apparent area of contact.[2] Material removal during adhesive wear is due to the shearing of junctions formed during contact, whereas abrasive wear results from the ploughing of softer materials by hard particles.

In contrast to adhesion and abrasion, the theory of fracture wear is relatively new. According to this theory material removal occurs in elastic-plastic solids, through the formation of platelets or wear particles which are later pulled from the surface. These wear particles are thought to be formed from the intersection of sub-surface cracks, propagated due to tensile stresses created from shear forces produced during sliding contact.[3] This type of wear may be more prevalent in materials, such as ceramics, with low ratios of toughness to hardness.

Ceramics possess many material properties which are advantageous for friction and wear application. Ceramics have lower forces of adhesion and higher hardness than metals, and therefore exhibit smaller friction coefficients and less abrasion by SiO_2 and other oxide particles than do metals. The reduced friction and abrasion of ceramics, when used in engine component applications may result in improved engine performance and fuel economy.

In this paper we briefly describe some automotive related laboratory tests and current wear and friction studies using ceramic materials. A discussion is presented assessing the tribological behaviour of ceramics under various conditions.

ENGINE COMPONENT TESTING

Friction force is heavily influenced by lubricating conditions. As observed by Staron and Willermet[4], friction losses in an engine valve train are predominantly due to boundary and mixed lubrication conditions. In applications where the lubricating film fails to separ-

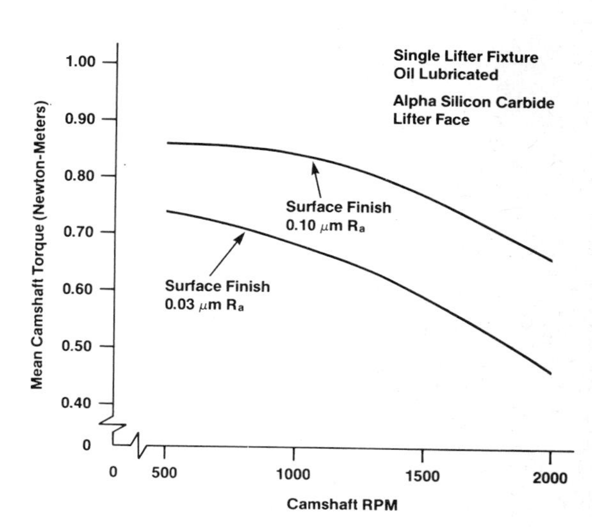

Fig 1 Surface roughness effects of camshaft torque

ate the sliding surfaces, interaction of the asperities produces increased friction.

Tests on sintered alpha silicon carbide (SASC) engine components confirm that reduced roughness is important for low frictional torque as illustrated in Figure 1. The mean camshaft friction torque, as measured by a single lifter test fixture under lubricated conditions, is reduced by up to 30% when SASC surface roughness decreases from 0.10μm to 0.03μm Ra. The lower torque can be attributed to reduced surface asperity interaction.

A reduction in friction was also observed with SASC wear pads in a valve train evaluation in an eight hour dynamometer test using an OHC 2-Liter Opel diesel valve train. A production four cylinder head assembly was used with the overhead camshaft directly coupled to a dynamometer. Friction torque was measured over a speed range of 300 to 2300 rpm (camshaft speed). The rocker arm wear pad received normal lubrication, and both oil and coolant were heated to typical operating temperatures. The use of SASC reduced motoring friction by 8 to 12% over the production valve train and 5 to 10% over Si_3N_4 pads.

Previous evaluations of sintered alpha silicon carbide's wear resistance, by Ricardo and several engine manufacturers, have shown very promising results. Tests on SASC valve lifters used in light to moderate running conditions, such as the Bedford truck engine test for 5000 hours and a Chevrolet 350 cubic inch engine test at 4000 rpm for 100 hours, produced negligible SASC wear. However, under severe running conditions which produce high hertzian stresses, the use of SASC may be limited due to its brittleness.

In a more recent field test of SASC rocker arm wear pads, which were subjected to 9000 start cycles, superior wear resistant behaviour (<0.13μm) was observed over the standard metal pads. Only minor pitting was observed for SASC (Figure 2) as compared with the wear and scuffing produced on the metal pad. The influence of pitting on wear performance, such as in valve train applications, is not fully understood.

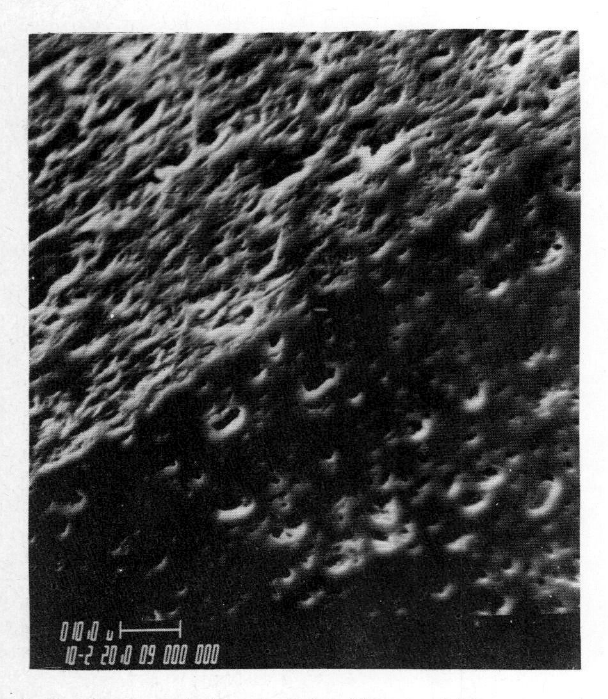

Fig 2 Surface pitting on alpha silicon carbide wear pads

A comprehensive analysis of the wear and friction behaviour of engine components is difficult due to the ambiguous operating conditions such as fluctuating load, changes in contact area and film thickness, and contamination of the environment. Controlled laboratory testing is important to provide some necessary understanding. Further insight into the field observations is provided through the following mechanistic studies performed in the laboratory on SASC mechanical face seals.

LABORATORY TESTING

Materials

The materials used for wear and friction testing are sintered alpha silicon carbide (SASC) and tungsten carbide. SASC is a single phase polycrystalline material with an equiaxed fine grained microstructure (5 to 8μm) containing no free silicon. SASC exhibits excellent corrosion resistance, high temperature strength retention, low friction, and excellent wear resistance. It has high thermal conductivity and can be polished to a very smooth surface (<0.13μm Ra). It has high hardness but relatively low roughness.

Tungsten carbide is a dense multiphase material containing a metallic binder. It exhibits good wear and friction properties but has limited corrosion resistance. It has a lower hardness than SASC but exhibits increased toughness. Selected properties of these materials are shown in Table I.

Table I - Properties of Test Materials

Material	Density (gm/cm^3)	Hardness (GPa)	Toughness ($MPa\sqrt{m}$)
SiC	3.16	24.5	3.79
WC	15.0	17.0	9.59

Experimental Procedures

Lubricated and dry sliding wear tests were conducted by using one inch diameter seal samples under unidirectional sliding conditions. All tests used a rotating seal face loaded against a stationary seal seat. Samples were hand lapped flat to within two helium light bands (0.5μm) and were characterized through surface profilometry. For lubricated testing, frictional torque was monitored following a break-in period as outlined in the literature[5,6] for velocities ranging from 0.4 to 7.1 m/sec, with the sample immersed in pressurized water. The test apparatus was similar to that of Labus[5], where the test vessel was vertically mounted on a torque table and the seal face driven by a variable speed motor. The dry sliding experiments were conducted on a Falex LFW-6 wear test machine. Tests were run at constant velocities using loads of 10 to 90 N and sliding distances up to 54 km. In addition to surface profilometry, wear surfaces were characterized extensively by means of optical and scanning electron microscopy.

RESULTS AND DISCUSSION

Preliminary experiments using SASC samples under lubricated sliding conditions gave diverse and ambiguous results. These tests were conducted using seals with two different surface histories, newly prepared seals, and previously tested resurfaced seals. Since the effect of surface finish on sliding friction was not understood, an investigation was conducted at various sliding velocities to evaluate changes in surface roughness as a function of running time. It was found that the contact surfaces became substantially smoother after extended test periods (Figure 3).

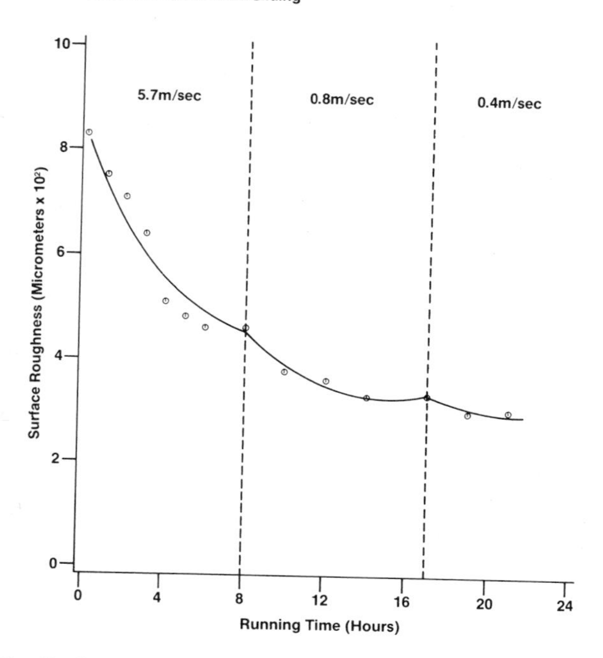

Fig 3 Surface roughness reduction as a function of running time

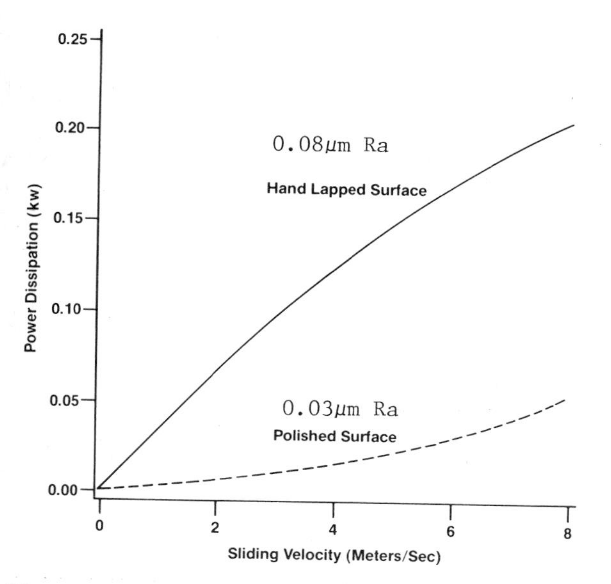

Fig 4 Power dissipation versus sliding velocity for various surface finishes

In another test, samples with polished surfaces were found to produce lower power dissipation, (the product of frictional torque and sliding velocity), than samples with hand lapped surfaces (Figure 4).

An analysis of the preliminary test results concluded that previously run resurfaced seals gave lower power dissipation and required less break-in time than newly prepared seals.

When testing was modified to use extended break-in periods, smooth contact surfaces and consistently lower power dissipation results were obtained for both seal types.

During the lubricated sliding tests it is believed that asperity shearing was responsible for reductions in the contact surface roughness. Smooth surfaces create fewer interactions and produce enhanced fluid film separation and lower power dissipation. The erratic frictional torque readings and deviations in power dissipation measured during the preliminary tests were caused by varying degrees of surface asperity interactions.

For some applications the surface roughness of metal components is purposely increased to improve their performance. This increased roughness yields lower adhesion due to a decrease in the true area of contact. The low shear strength of a metal can cause rough surfaces to "wear-in" easily during break-in. Unfortunately this practice of purposely roughening the surface is sometimes applied to ceramic materials without careful consideration of ceramic deformation and fracture characteristics. The high shear strengths and low forces of adhesion of a ceramic cause its mechanical response to fluid film disruption and asperity interaction to be more pronounced than for a metal. A ceramic responds to asperity interaction by fracture or elastic deformation while a metal may deform plastically. Due to these dissimilar characteristics, ceramics may require longer break-in times than metals with similar surface finishes.

Results of dry sliding tests with SASC sliding against itself show that various wear mechanisms may become dominant at different loads. Weight losses were monitored as functions of sliding distances, and were normalized with respect to time and contact area for each load tested. Resulting wear rates, as can be seen in Figure 5, follow an exponential dependence on load. Scanning electron microscopy observations

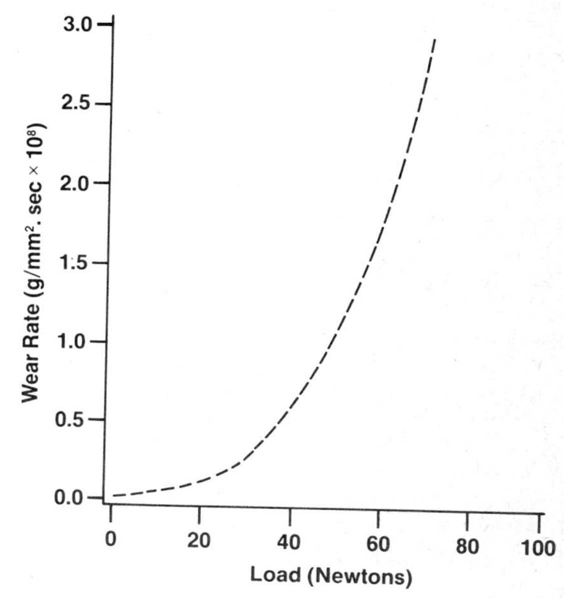

Fig 5 Dry sliding of SiC versus SiC

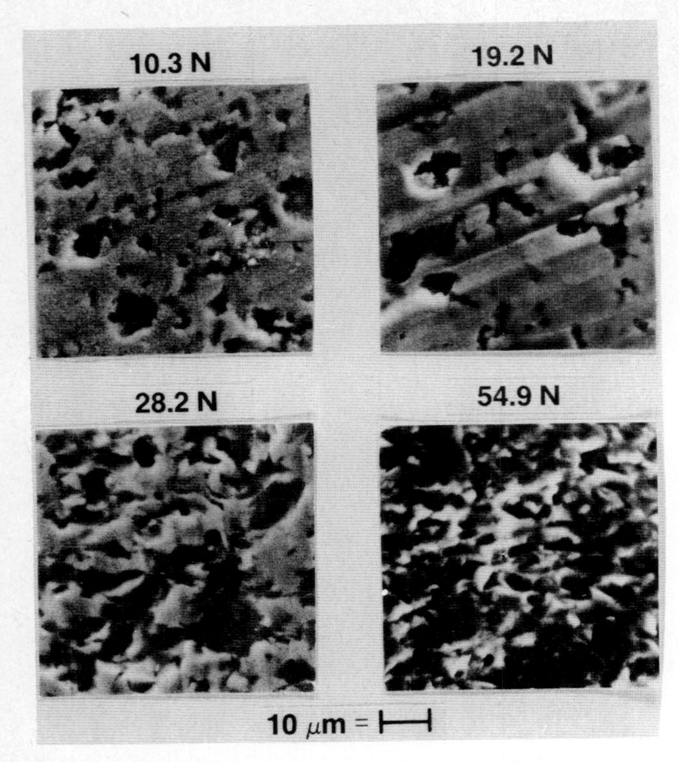

Fig 6 Changes in surface topography occurring at various loads on SiC

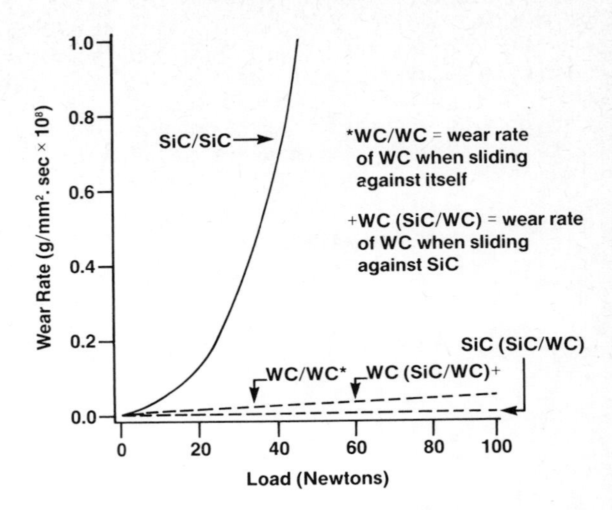

Fig 7 Dry sliding of SiC versus SiC, WC versus WC and SiC versus WC

suggest that distinct surface topographies occur at various loads which are characteristic of the controlling wear mechanism (Figure 6). At low loads smooth surfaces are created which exhibit permanent deformation. In this regime abrasive wear mechanisms are dominant. With high loads however, rough surfaces are generated due to brittle fracture. Under these conditions, sub-surface tensile stresses are created which surpass the critical stress required for crack propagation.

As stated, brittle fracture occurs when SASC slides against itself at high load. The general practice, however, dictates the use of dissimilar material combinations to achieve minimum wear. This practice is supported by wear test results of SASC sliding against tungsten carbide. Under these conditions wear rates are two orders of magnitude less than those previously observed. Low rates were observed for both materials, yet as shown in Figure 7 SASC exhibits less wear than tungsten carbide. Wear track characterization indicates the absence of wear by brittle fracture in either material. It is unclear why crack propagation is impeded in this situation.

The above wear tests give valuable information on the wear behaviour of SASC under pure sliding conditions. However surface pitting, observed in the engine component tests, was not observed in these sliding tests. This result may be explained by the presence of combined impact and sliding conditions inherent in valve lifter and rocker arm applications. The initial contact between the cam and SASC wear pad resembles particle impingement observed for erosive wear. The total pitting volume is probably influenced by hertzian load, stress intensity factor, hardness, relative velocity, area of contact, and number of cycles. Additional studies are required to unequivocally explain the mechanisms in this pitting wear.

CONCLUSIONS

It is shown that the wear and friction behaviour of alpha silicon carbide is significantly influenced by (1) surface roughness for lubricated conditions and, (2) material combinations and load under dry sliding conditions. This study shows that extended break-in periods reduce the surface roughness and decrease power dissipation for lubricated conditions. However, it should be recognized that significant changes in wear and friction behaviour also occur with time under dry sliding conditions.

Observations from engine component testing correlate with wear and friction results obtained under laboratory conditions. Application of sintered alpha silicon carbide may be limited for engine components subject to high hertzian or impact loads due to the possibility of brittle fracture. For those applications where moderate load conditions exist, such as valve guides or cylinder liners, alpha silicon carbide offers advantages over metal components for wear and friction reduction under poor lubrication conditions.

REFERENCES

(1) BUCKLEY, D. H. Importance and definition of tribology - status of understanding. Tribology in the 80's. NASA Conference - 2300, 1983.

(2) RABINOWICZ, E. Friction and wear of materials. T. Wiley & Sons, 1965.

(3) SUH, N. Tribophysics. To be published.

(4) STARON, J.T., WILLERMET, P.A. An analysis of valve train friction in terms of lubrication principles, SAE Paper No. 830165, SAE International Congress and Exposition, Detroit, Michigan, February 28-March 4, 1983.

(5) LABUS, T.J. The influence of rubbing materials and operating condition on the power dissipated by mechanical seals, ASLE, Preprint 80-AM-6B-3, 1980.

(6) DERBY, J., SESHADRI, S. G., SRINIVASAN, M. Microstructural considerations on the sliding friction characterstics of ceramic seals. To be published.

A study of some cylinder liner wear problems, following the introduction of high speed trains on British Rail

T S EYRE, BSc, MSc(Eng), PhD, CEng, FIM
Department of Materials Technology, Brunel University, Middlesex

INTRODUCTION

When the first of the power cars of the high speed trains were beginning to come in for their first inspections and overhauls it was observed that the wear rate on the chromium plated cylinder liners was rather variable. There was therefore a need to discover the reason for this variability to enable the performance of all the liners to be raised to equal that of the best, so that overhaul lives could be consistently extended.

A metallurgical investigation of the problem was carried out at Brunel University, with close cooperation from British Rail and Paxman Diesels and guidance from Michael Neale and Associates as consultants. The investigation was of considerable technical interest as well as being an excellent example of a joint approach leading to a successful conclusion.

British Rail monitored the engines by spectrographic oil analysis by taking oil samples every 3 to 7 days and checking these for chromium, iron, and other metals. They were therefore able to identify engines which showed signs of high wear and check these by measuring any wear step at the top of the travel of the piston rings. It was therefore possible to identify specimen engines showing high and low wear for which components could be collected for comparative examination and for checking against new components supplied by the manufacturers.

EXAMINATION OF WORN COMPONENTS

Towards the end of August 1979, two engines became available at Derby for classified repair which showed relatively high and relatively low wear respectively. Two matched liners, pistons and rings were collected from each engine, together with all the rings from other pistons and samples of the lubricating oil. Specimens of a new liner and set of rings were also provided by Paxman. All components were subject to a general examination and then one liner from the engine with high wear with its associated rings was subject to a much more detailed study.

General observations

The cylinder liners are made of steel, plated in the bore with hard chromium and then diamond honed to bring the bore to size and to produce a crazed finish by a small scale fracturing of the surface as plating tensile stresses are relieved. This is then followed by silicon carbide honing to produce an improved finish with appropriate plateau area. In all but the upper 3.75" of the operating surface, the bore is then back etched through a perforated mask to produce a dispersed pattern of pits, which is intended to give further assistance to oil distribution, in addition to that already available from the grooves in the surface crazing.

The pistons are of aluminium alloy with an insert for the top ring groove, and are fitted with three compression rings and one oil control ring, all above the gudgeon pin. The compression rings are of cast iron and are copper plated all over when new. The top ring is plain and the 2nd and 3rd rings are stepped at the back so that when fitted they distort to add to a machined downscraping taper face of 1.0^{o} to 1.5^{o} in the free condition. The oil control ring has two lands about .025" wide and is pressed against the cylinder wall by an internal peripheral coil spring.

All the cylinders showed wear marks running up and down the bores with a main wear step at the TDC position of the top compression ring and a minor wear step corresponding to the BDC position of the oil control ring.

To provide a permanent record of the liners in the as received condition, and to facilitate microscopic examination, replicas were taken from all liners. The replicas were then examined to determine the wear mechanisms involved and Figure 1 summarises the results for the area of major interest which is the top 6 cms of the liner down from the TDC position of the top ring. It can be seen from this that the liners all have broadly the same characteristics with minor variations. The original honing pattern, which is fully in evidence around the mid stroke position is progressively removed by an abrasive wear process and merges into an abrasively polished area which, depending upon the amount of wear present, extends from the top to distances of 2cm to 5cm. This polished area corresponds to a region with little evidence of surface pits, associated with the wear process having progressed below the depth of the main surface porosity. Once the surface has become polished in this way it then appears prone to scuffing in localised areas as indicated in Figure 1.

The pistons were generally clean on their skirts but showed carbon deposit on the surfaces above the second or their compression rings with some light axial scoring of the carbon layer on the top land.

The piston rings had lost all their original copper coating from all faces other than the non contacting portions of the running faces of the 2nd and 3rd compression rings, that is over the extreme top 0.2 to 0.25 of their width. The other surfaces were all polished or worn through to the base cast iron material and the running faces had a general appearance indicating abrasive wear.

Some top, 2nd and 3rd rings also, however, showed some superimposed bands of scuffing which were about 1mm to 2mm wide. On the worst rings there were 10 to 15 such bands, with a tendency for the scuffing to be more widespread near the horns.

This general examination indicates that the liner and rings in column 1 of Figure 1 is probably the most typical high wear liner and shows all the various wear characteristics in a fully developed form. This assembly was, therefore, selected for more detailed study, as described below.

Detailed study of one liner and associated rings

After taking surface replicas the liner was cut into axial strips about 1.5" wide, taking care to avoid damage to the running surface. This gave improved access to the surface for examination and simplified the preparation of specimens.

Examination of the strip of liner with the naked eye and with a low power binocular microscope clearly showed the tendency for the liner surface to become polished at the upper end and also, although to a reduced extent, at the BDC position of the ring pack. The marks corresponding to the stationary position of each ring at the dead centres could be clearly seen and it was also noticeable that the top ring produced a small double step at the TDC position with the second step just under 1mm long. This is almost certainly due to the piston travelling a small distance further up the bore at TDC exhaust/induction than it does at TDC compression/firing due to deflections and the taking up of clearances in opposite directions on the two strokes due to the application of gas pressure forces on the piston.

One strip of the liner was subject to very detailed examination to obtain a continuous picture of the surface in the top 1mm, both as surface photographs with a scanning electron microscope and as metallographic sections through the material. In addition a series of photographs was taken at critical positions along the whole length of the liner with a scanning electron microscope to record all the various kinds of surface texture available. A series of selected photographs from this sequence with a guide to the regions where they were taken is shown in Figure 2 and 3. It can be seen from this that the original crazed finish is still present at the low wear regions in midstroke and is progressively removed by a fine abrasive wear mechanism until a polished surface is produced towards top dead centre. There is also a related polishing effect, but to a much reduced extent, associated with the oil control ring around the bottom dead centre position.

The polished area towards TDC is in fact extremely smooth and a Talysurf trace of this area is shown in Figure 4. It corresponds to a surface finish of 2 microinches cla. In these very smooth areas there is also evidence of local scuffing of the chromium surface characterised by smearing of the chromium across fine residual surface cracks and some associated delamination as shown in Figure 5.

In all the very detailed study of the surface of this liner, which was generally typical of all the liners examined, there was no evidence of any surface characteristics on the liner which could in any way be related to corrosion.

The main wear mechanism is fine abrasion resulting in the surface porosity in the chromium becoming worn away and producing a very finely polished surface, which is then prone at this later stage to some local scuffing.

The set of piston rings which had operated with this liner were also examined in detail. All showed some evidence of abrasive wear and some narrow bands of scuffing. Figure 6 shows one of the scuffing bands on the top compression ring at two levels of magnification, together with a microsection which shows white layers present in the surface, and confirms the existence of a local metal to metal rubbing situation.

Figure 7 shows cross sections through the second ring, which is taper faced, at two levels of magnification together with a direct view of the running face.

The upper photograph shows the region where the remaining copper plate on the upper portion of the running face merges into the cast iron surface which is operating against the cylinder liner. The middle photograph shows this copper plated area at a greater magnification and it can be seen that there are a number of small light coloured reflective particles embedded in the copper. These show even more clearly in the lower direct view of the copper plated ring surface where it can also be seen, that they correlated broadly with the abrasive wear lines on the face of the ring.

Figure 8 shows a further direct view of the same area taken in a scanning electron microscope with corresponding microprobe analyser photographs for the same area showing chromium, copper and iron. These show very clearly that the particles embedded in the copper are pieces of chromium.

These particles were present on both the second and third rings and there were a very large number of them, so that each ring would have several hundred particles distributed around its circumference. These particles were mostly embedded in the copper close to the edge adjoining the cast iron surface, and were not evenly distributed up the copper face. It was possible to find a few particles embedded in the iron but these were outnumbered by at lease 100 to 1 by those in the copper. To check that these particles could not have arisen from any chromium rich constituents in the cast iron of the rings a portion of the ring was also checked by microprobe analysis and proved to be negative. This proves conclusively that the particles could not have come from the rings and that they must therefore have come from the cylinder liner. The chromium particles ranged in size up to 50μm.

The various cylinder liners, rings and piston ring grooves were also washed out with solvents and any separated solids were subject to X-ray diffraction and microprobe analysis. This debris was mainly iron and calcium sulphate with only small traces of chromium.

Various samples of used oil were also analysed and centrifugally separated. The analyses showed that the oil had not deteriorated or picked up very much acid suggesting that corrosion was unlikely to have occurred. Centrifugal separation did not reveal any chromium particles but spectrographic analysis detected the presence of chromium in broadly the same proportions as had been found by British Rail in their spectrographic analyses of the same oils.

The detailed examinations and analyses all indicate that the critical wear mechanism in the engine is abrasion, associated with the presence of large numbers of small hard particles of chromium, originating in the cylinder liner and becoming embedded in the copper coating of the piston rings and therefore trapped between the running face of the rings and the liner. This abrasion wears away the textured surface of the cylinder liner and results eventually in a smooth polished surface on the chromium which is then prone to scuffing.

It is therefore important to determine how these hard particles of chromium come to be released by the cylinder liner, and as a guide to this it is relevant to try and discover when they are released in the life cycle of the engine. To assist in this aspect the next section of this report examines the available evidence on how wear rates vary during engine life, as indicated by spectrographic analysis of oil from engines in service.

STUDY OF THE OIL ANALYSIS RESULTS AVAILABLE FROM BRITISH RAIL

British Rail were taking lubricating oil samples from all High Speed Train engines at approximately weekly intervals, to check the oil condition, and to analyse spectrographically for traces of metal which can indicate wear occurring in the engines. The metals analysed for indications of wear were iron, lead, silicon, chromium, aluminium, tin and copper, and in addition sodium was used as an indicator of coolant leaks. The results for these elements were presented as parts per million in the oil.

Chromium is the main indicator of cylinder liner wear, and iron levels are also relevant as among other things they can indicate piston ring wear. A chromium level of more than 30 ppm was used as an alerting level and if it rose consistently to over 50 ppm an inspection of the cylinders was generally initiated. This technique has been found effective for detecting any engine problems.

Some relevant spectrographic analysis data was plotted graphically against engine operating time in service and is shown in Figure 9. The results are very approximate because a scatter between successive samples amounting to about 30% of the ppm value was found in all cases, and the lines on Figure 9 represent a mean through the various clusters of measured values.

The results in Figure 9 do however clearly indicate that there is a peak value in the generation of detectable chromium particles which starts to build up as soon as the engine enters service. This peak value is usually reached after one to four months and the levels then fall away and tend to become steady for a time at a lower level. This level may then subsequently increase towards the end of the engine life.

When considering the results shown in Figure 9 it must also be remembered that they only relate to chromium particles less than 10 microns in size and at various stages in the operating life particles larger than this might be being generated but not recorded. It is probable however that the main progressive material removal process, leading to surface polishing, almost certainly involves the generation of detectable particles, and the ppm values for chromium probably therefore can be taken as a guide to the liner wear rate occurring at any time. The values do in fact represent a wear rate rather than an accumulated amount of wear, which might be assumed at first sight. They indicate wear rate because the particle concentration is subject to a steady dilution as a result of settling, filtration, oil consumption and top up, so that the ppm level at any time represents a rate balance between generation and removal.

The results in Figure 9 indicate that the highest liner wear rates occur early in the running life of the engines which suggest that there is either a problem with the running in process or with the as new condition of the components of the cylinder assembly. Even though the curves in Figure 9 are not complete they appear to be part of a picture in which engines, which have experienced a high initial wear rate, never fully recover. Engines with a very small initial wear rate appear to be capable of surviving to long operating times with relatively low wear while other engines which have had higher peaks initially show much higher wear in later life.

STUDY OF NEW LINERS AND PISTON RINGS

Samples of new cylinder liners and piston rings as currently fitted to the engines were obtained for examination in order to discover whether there were any features which might account for the problems which seem to occur in the early stages of engine running.

All the compression rings are coated with copper which has a thickness of about 6 microns on the running face and about 4 to 7 microns on the upper and lower surfaces. The rings are of cast iron with a graphite/ carbide/pearlite structure and with D/E graphite in the size range 5/6. The hardness of the cast iron is in the range 290-305 Hv. These rings are typical of good quality piston rings and show no features which might in themselves give rise to a high rate of cylinder liner wear when operated with chromium plated liners.

The cylinder liners in the as received condition were washed with solvent and surface replicas were taken before cutting them up into strips for detailed study. This provided a reference condition which could be used to ensure that any observed features had not been caused by specimen preparation.

Figure 10 shows a strip of new cylinder liner alongside a strip from a used liner to indicate the region over which the rings operate. It can be seen that the bore of the liner is covered with chromium over its whole length with a matt cross honed finish. Over all but the top 3.75" the liner surface has a uniformly distributed spot porosity obtained by etching the liner surface through a mask. This spot porosity is intended to assist oil retention and distribution and is not continued to the top of the liner in case it should give rise to a high oil consumption.

A detailed examination of the liner showed the chromium layer to be typical of normal electroplated chromium with a hardness in the range 800-875 Hv and with a thickness of about 250μm (0.010"). The chromium plate had been cross honed to produce surface porosity by dispersed local fracture of the chromium surface. This porosity appears as a matt finish to the naked eye but at higher magnification as shown in Figure 11 the local fracturing of the surface can be seen in greater detail. The surface porosity is of the order of 50% by area and its surface profile is indicated by the Talysurf trace included in Figure 11 which is to the same horizontal scale as the right-hand photograph, with which it correlates quite closely. The trace shows an acceptable plateau area in the surface and corresponds to a surface finish of 80 microinches cla. This appears to have been obtained by a final light honing process and the marks from this can be seen on the plateau areas in the surface photographs.

The most important discovery during the examination of the new cylinder liners was that when surface replicas were taken using solvent softened plastic films, particles of chromium came away from the liner surface and adhered to the plastic replica. This occurred in the replicas of the liners in both the as received and the cut up condition, and the particles which became detached were of the same order of size as those which have been found embedded in the copper plating on used piston rings. This indicates that there are a number of loose or lightly attached chromium particles in the liner surface.

To check this situation further surface photographs at increased magnification and microsections were taken from the new liner and also from the unworn portion of a used liner in the region above the TDC position of the rings. These are shown in Figure 12 and the loose particles can be seen clearly in both cases.

In a manufacturing process which involves the production of surface porosity in a hard material by a small scale fracture mechanism there are inherent risks that loose particles may be left in the surface or that particles may be left in the surface in a partially attached condition. The risk of this is reduced by using a light honing operation to finish the liners and usually this is followed by vapour blast cleaning. In the case of these particular liners however, it had been felt that vapour blasting might leave some abrasive material in the dispersed pit structure and this process had not therefore been used. The investigation had shown that problems were arising from presence of loose or lightly attached particles in the liner surface, and some modification to the manufacturing process were tried using vapour blasting at various stages and also using alternative honing treatments.

To check these liners and to provide a basis for a possible quality control check in the future a test procedure was devised based on a development of the method already used to examine the liner surfaces during the earlier part of this investigation. This procedure was based on the use of solvent softened plastic surface replica films pressed into the surface in order to pick up any loose chromium particles. These are then examined under a microscope and the number of particles along a 1cm line on the surface are counted and classified into size ranges.

This procedure was carried out on four types of cylinder liner:-

1. A standard liner as used initially.

2. A liner vapour blasted prior to the back etching treatment.

3. A liner vapour blasted at the end of the standard treatment i.e. after back etching.

4. A liner finished by a slurry honing process after the back etching.

Histograms showing the distribution of particles of various sizes on the different liners are shown in Figure 13. When it is remembered that the number of particles involved are those found along a 1cm line on the surface, it can be seen that substantial numbers of particles are involved, which is consistent with the hundreds of particles found embedded in the copper plating on the piston rings. It can also be seen that the slurry honing process represents a major improvement over the standard process and other alternatives. The vapour blasting which had been omitted from the end of the standard process does produce a noticeable improvement, but reverting to this process would not seem to give sufficient improvement to fully solve the problem. It also appears to be essential, if the process is used, to do it at the end of the other operations.

SUMMARY OF RESULTS

An examination of oil analysis measurements taken from early engines in service in the BR HST indicated that the variability was mainly associated with a wear mechanism which occurs early in the engine operating life. The metallurgical investigation has shown that this wear mechanism involves abrasive wear of the cylinder liners by particles of chromium, derived from the liners themselves, which embed in the copper plating on the piston rings to form an abrasive lap, and give rise to wear by becoming trapped between the cast iron running faces of the rings and the cylinder liners.

From the examination of used cylinder liners and piston rings and the analysis of oil samples there was no evidence that corrosion had played any significant part in the wear process. Essentially the wear process involved abrasive wear of the liners by loose chromium particles and, by the time the surface porosity on the liners has been worn away, the abrasive wear was operating as a fine abrasive process, leading to polishing of the chromium surface. It is general experience that polished chromium surfaces have poor tribological properties, which is the reason why proprietary surfacing processes are used, and in these engines once the surfaces become polished they were, as would be expected, prone to scuffing. This scuffing could then give higher wear rates towards the end of the overhaullife of the engines.

Examination of new cylinder liners had shown many loose particles in the surface of new cylinder liners. This indicated that changes were needed in the manufacturing process of the cylinder liners to ensure that the number of such loose particles is reduced to negligible proportions. However in addition to particles which were actually loose, there may have been others which were only weakly attached and the number of these can be reduced by further process refinements aimed at achieving a sharp transition between firmly attached lands of chromium and fully removed particles.

The incorporation of the necessary changes in the liner manufacturing process were made, and in addition the copper plating on the rings was eliminated to prevent any residual chromium particles from attaching themselves to the ring surface. These modifications have completely eliminated the variability in wear performance that was experienced on early engines and the very satisfactory wear rates now being achieved are an example of the success that can be achieved by a well coordinated cooperative study.

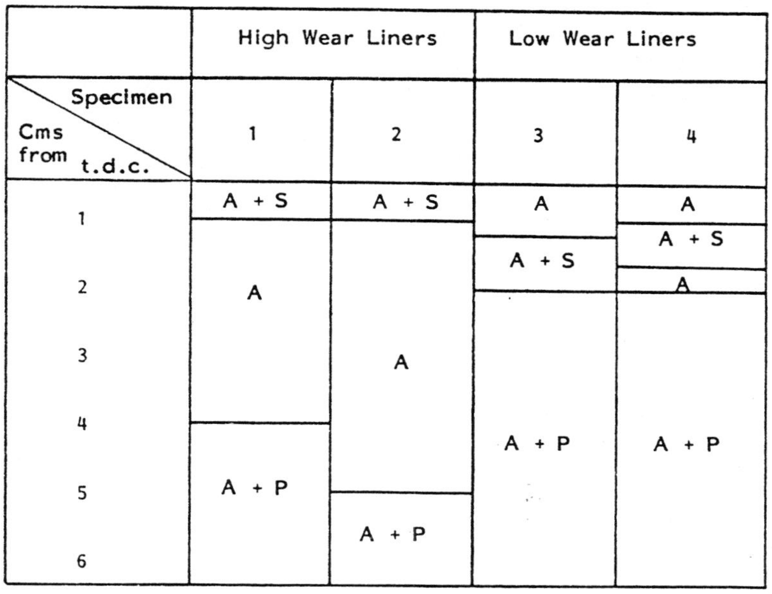

	High Wear Liners		Low Wear Liners	
Cms from t.d.c. \ Specimen	1	2	3	4
1	A + S	A + S	A	A
			A + S	A + S
2	A			A
3		A		
4			A + P	A + P
5	A + P			
		A + P		
6				

A = Abrasion

S = Scuffing

P = Original pits still evident in the surface

Fig 1 The condition of the top 6 cm of the operating surface of the liners indicating the various wear mechanisms observed

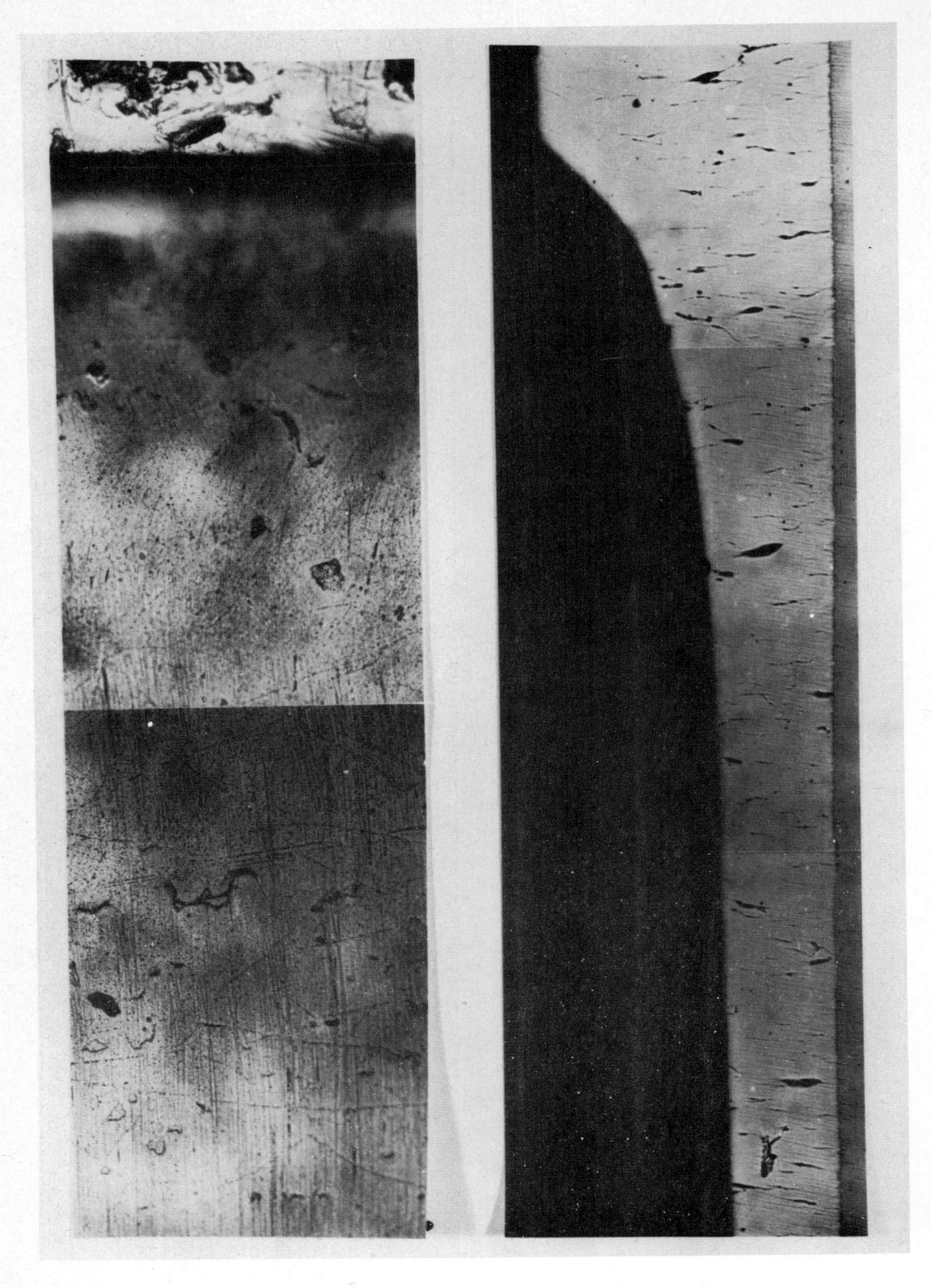

Fig 2 Direct and cross-sectional view of top 1 mm of worn region of cylinder liner

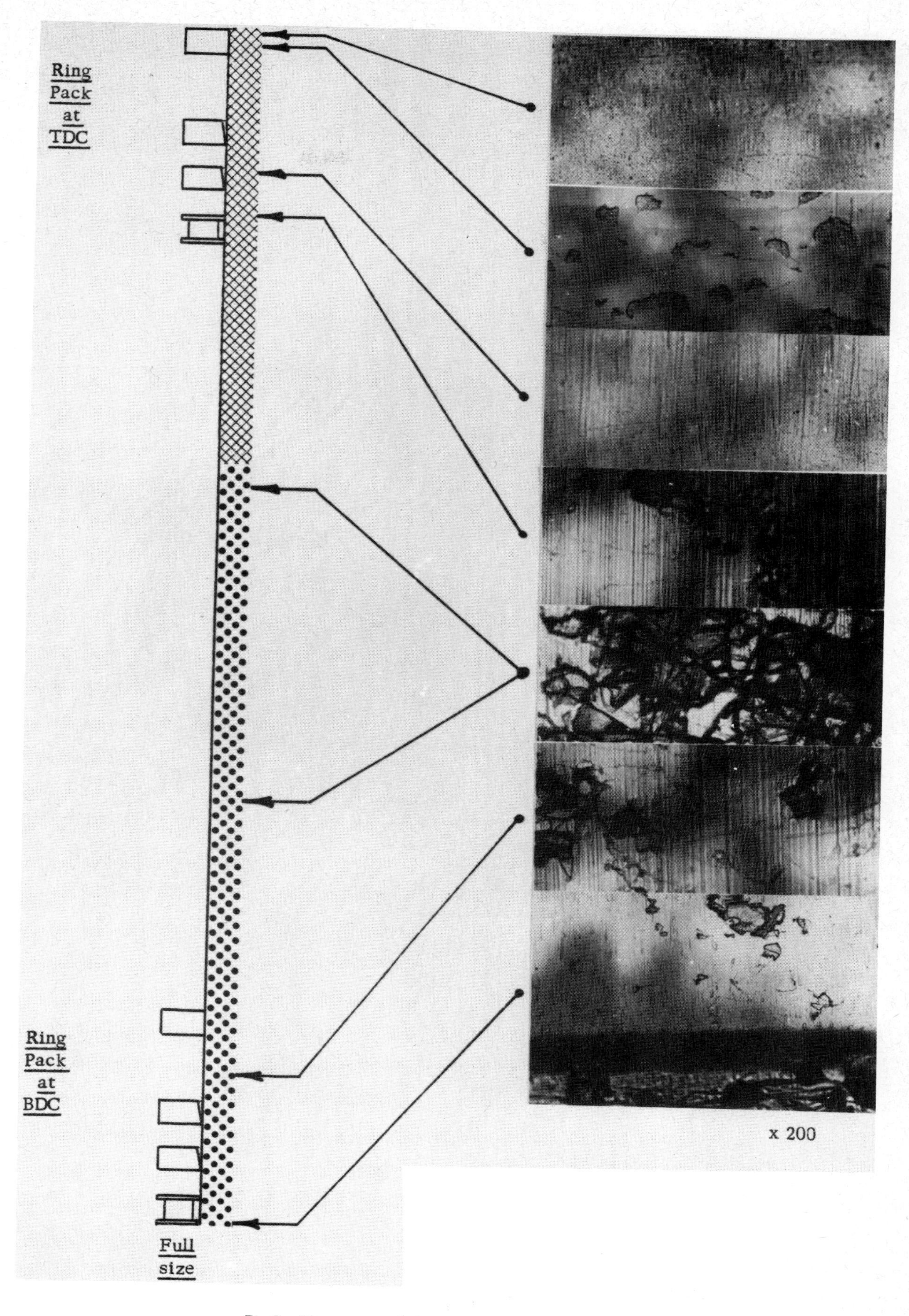

Fig 3 The nature of the liner surface at various positions

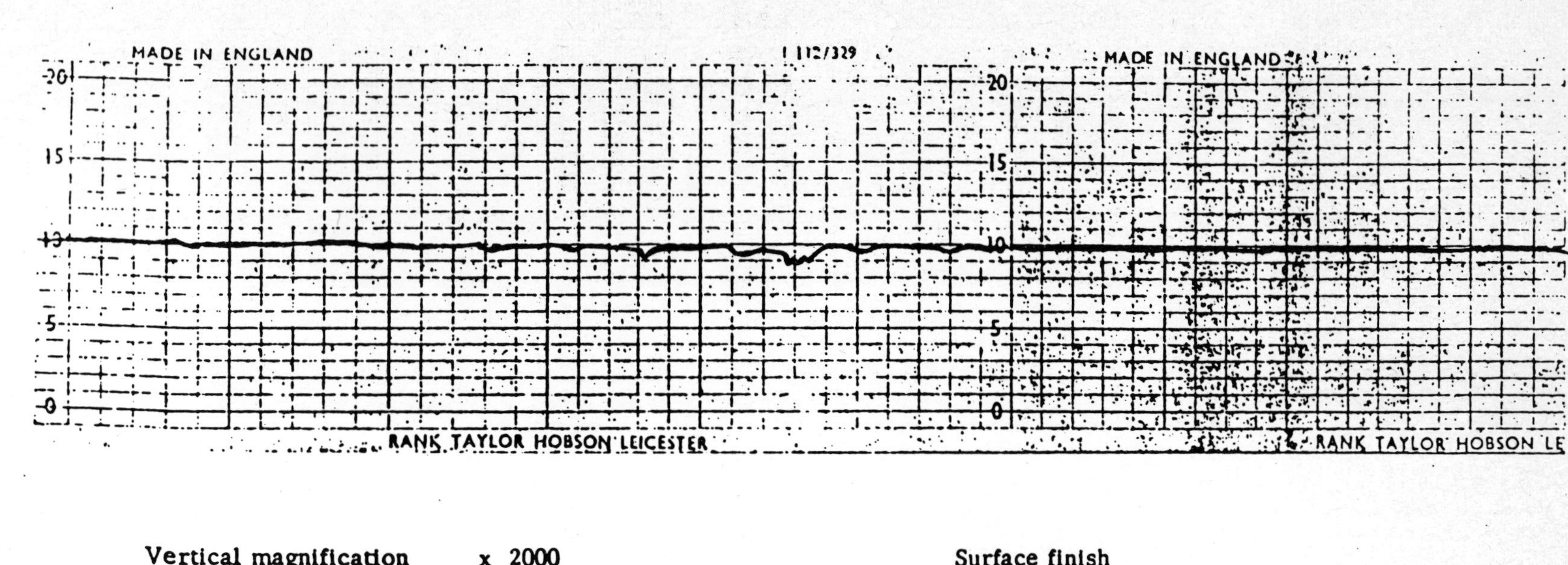

Vertical magnification	**x 2000**		**Surface finish**
Horizontal magnification	**x 100**		**2 microinches cla**

Fig 4 Talysurf trace of polished area towards the top of the liner

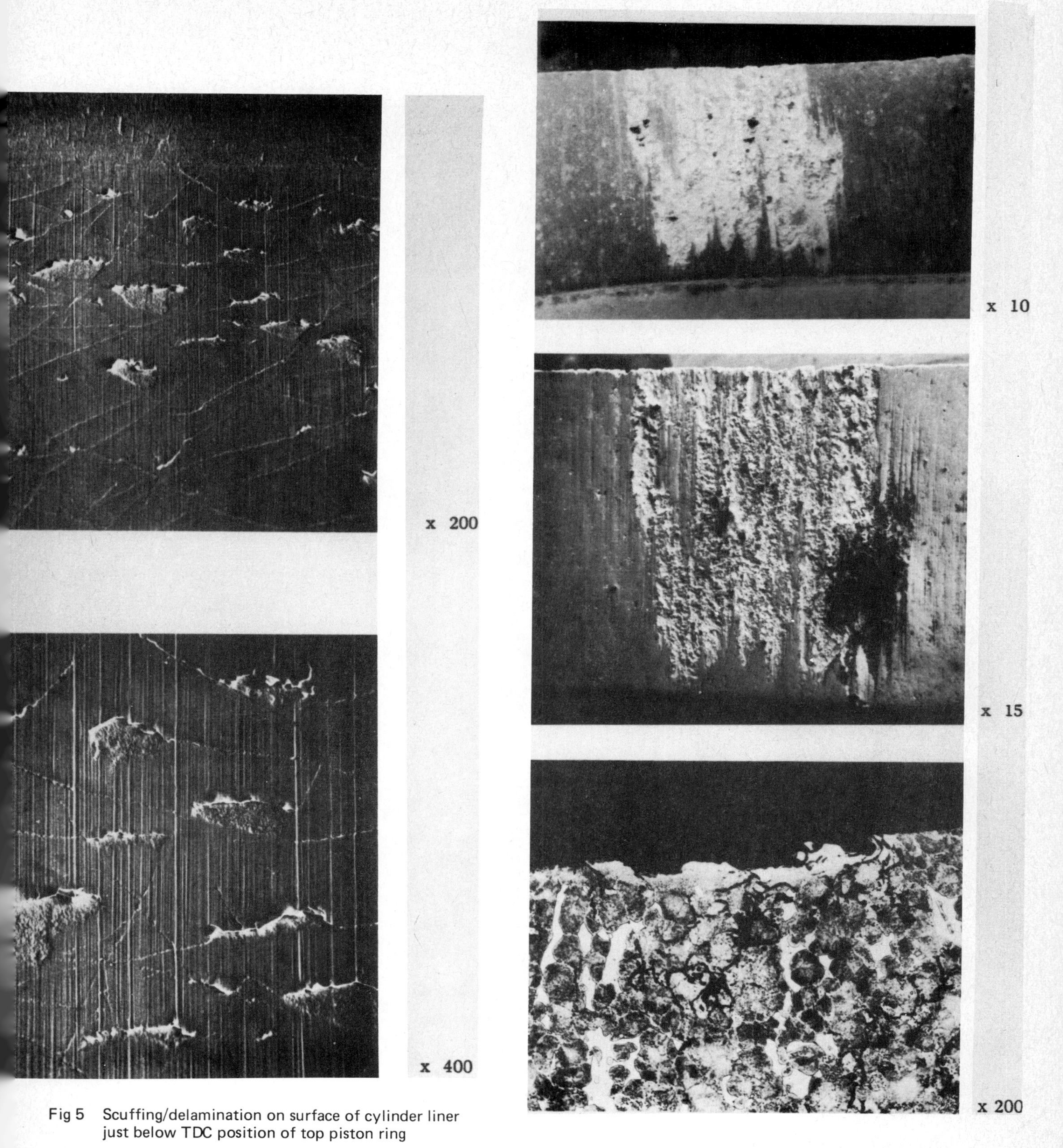

Fig 5 Scuffing/delamination on surface of cylinder liner just below TDC position of top piston ring

Fig 6 Scuffed areas on a piston ring

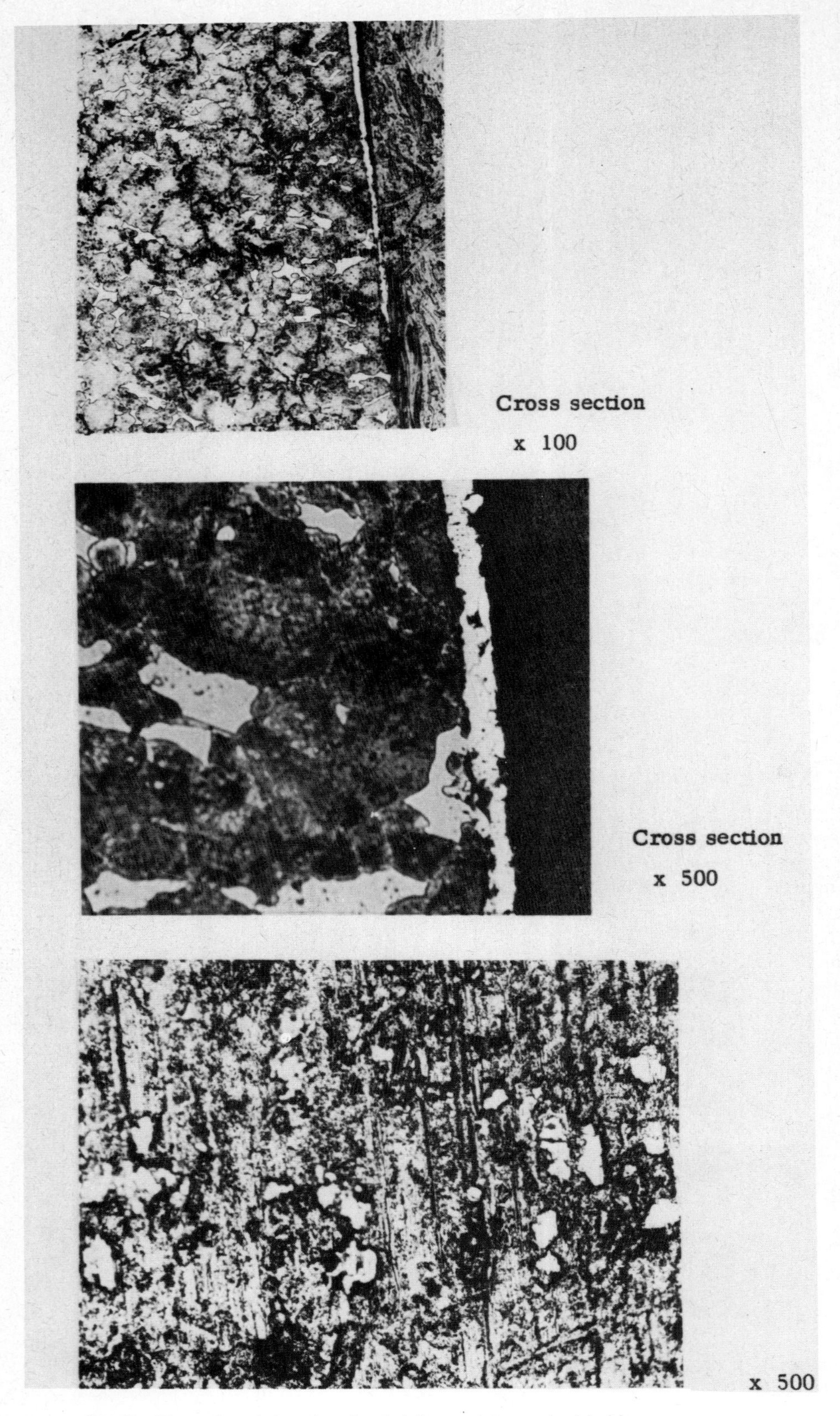

Fig 7 Taper faced ring showing bright particles embedded in copper plated surface

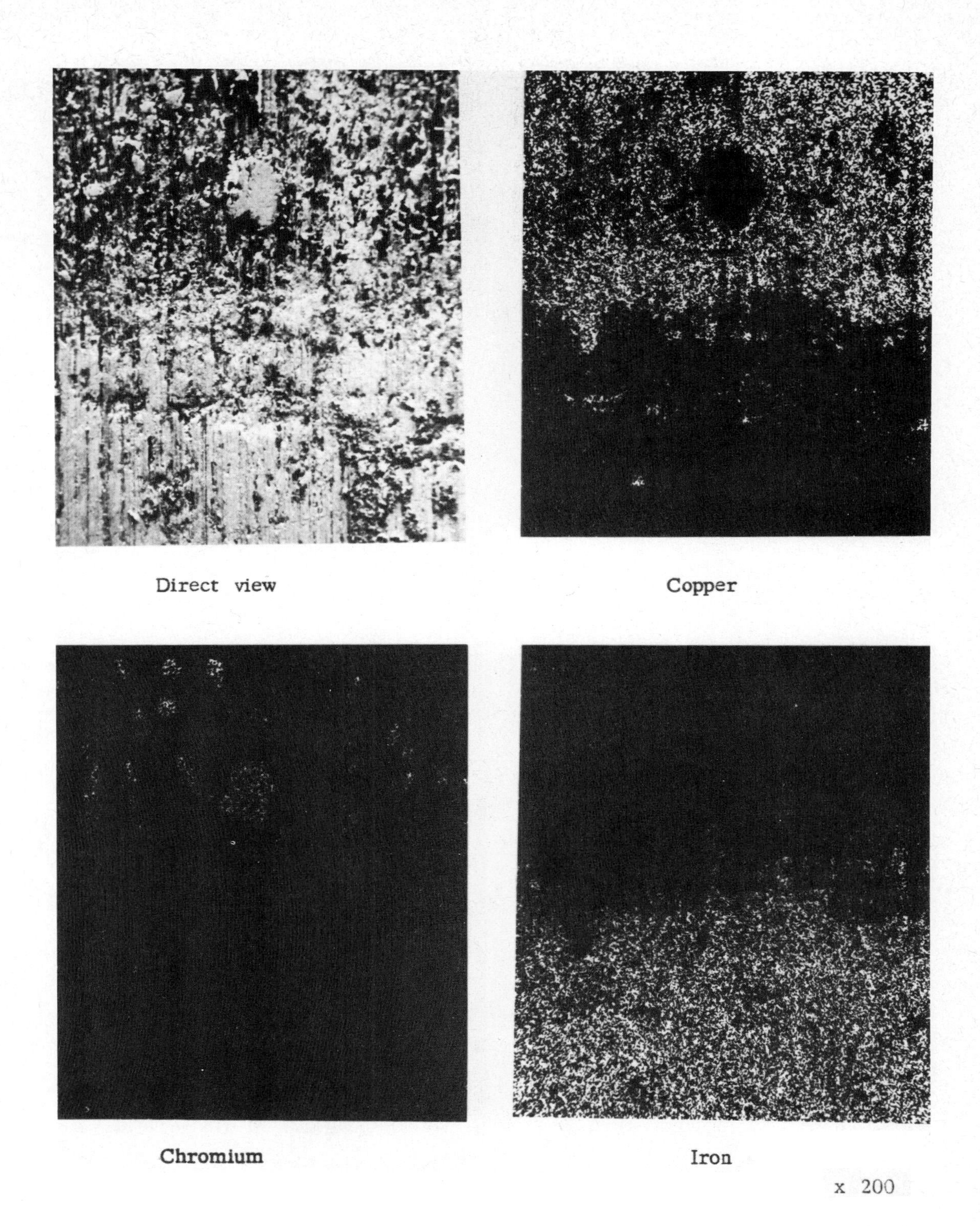

Fig 8 Running surface of a taper faced ring at lower edge of copper plated area with associated analyses

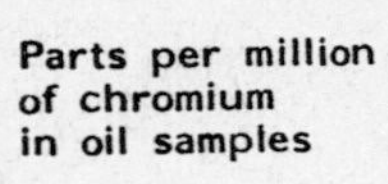

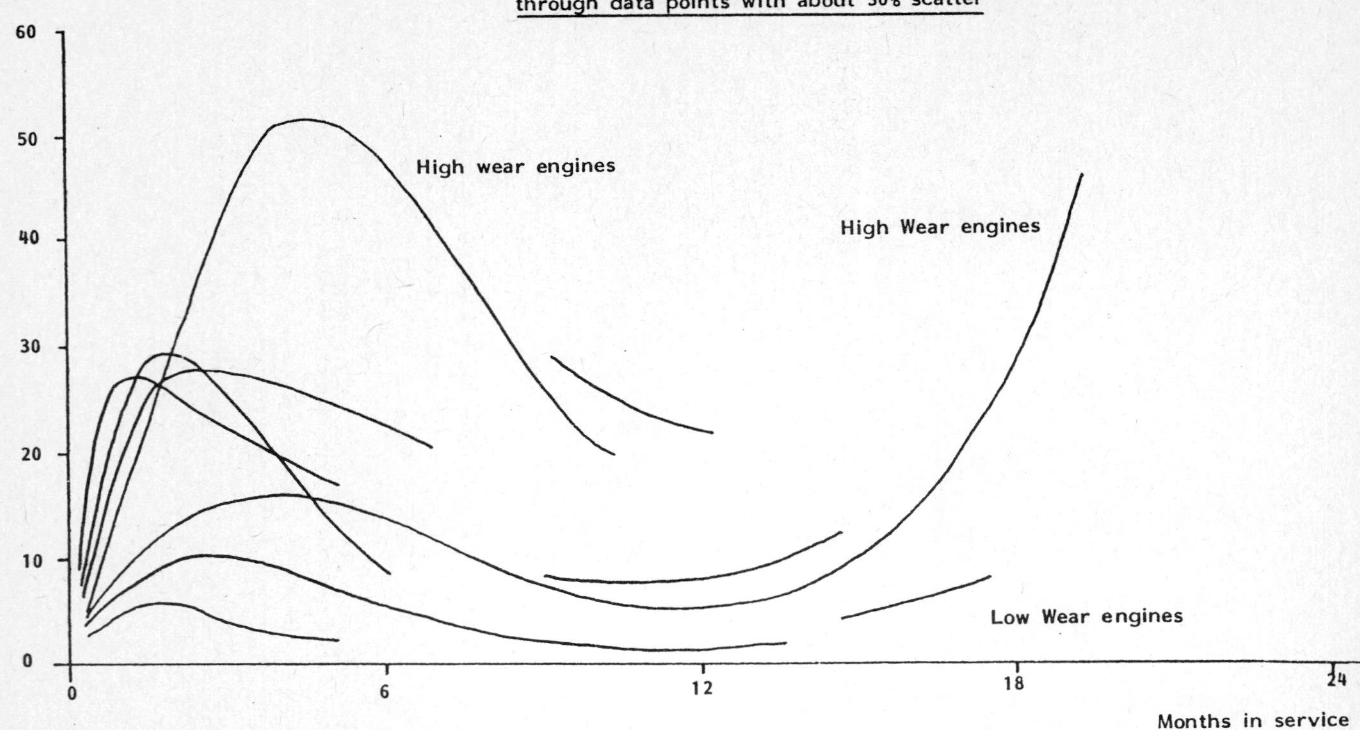

Fig 9 Variation of chromium content in lubricating oil during service life of engines with a known history

Fig 10 Specimen strips of new and used cylinder liners

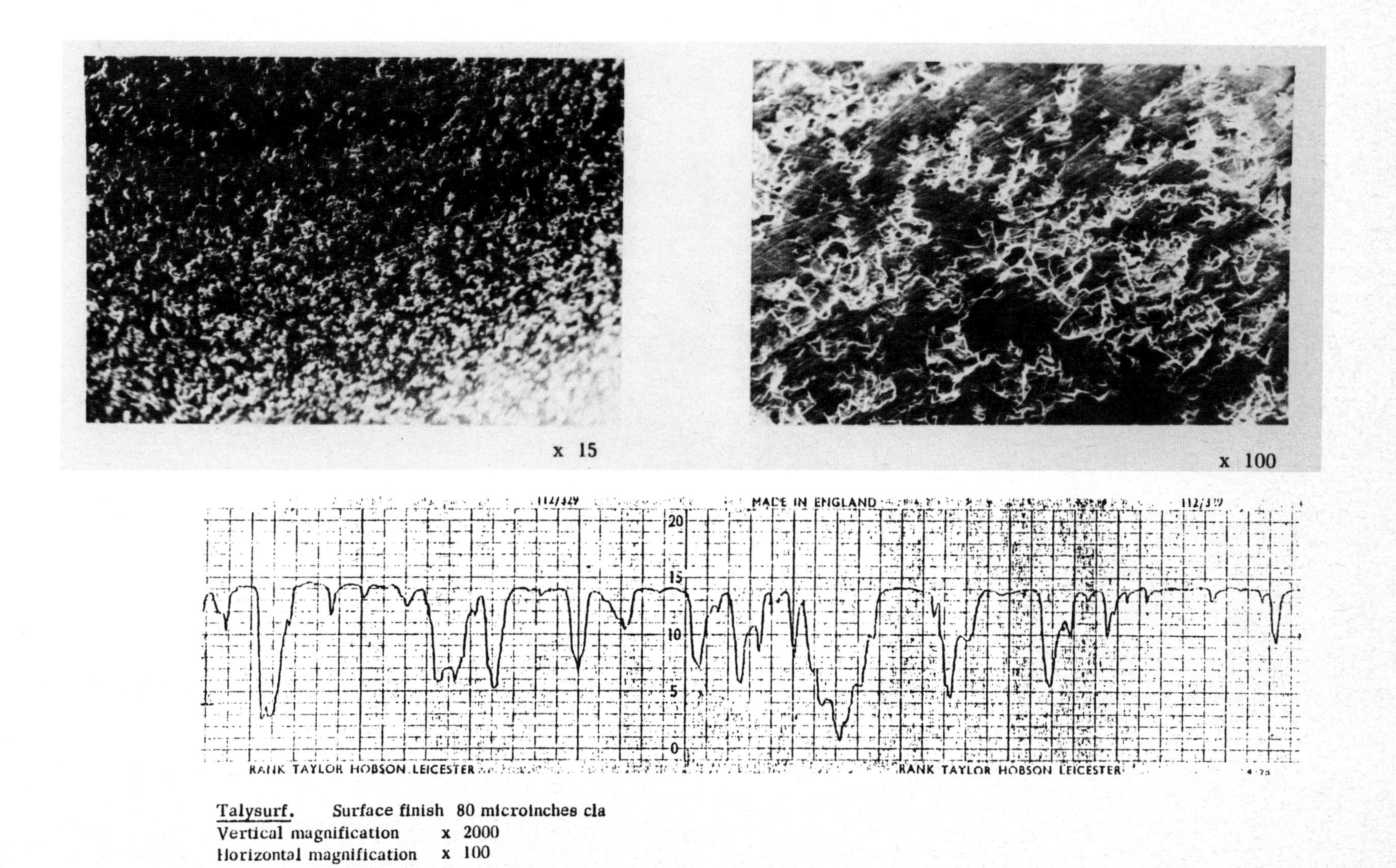

Talysurf. Surface finish 80 microinches cla
Vertical magnification x 2000
Horizontal magnification x 100

Fig 11 Surface of an unused cylinder liner

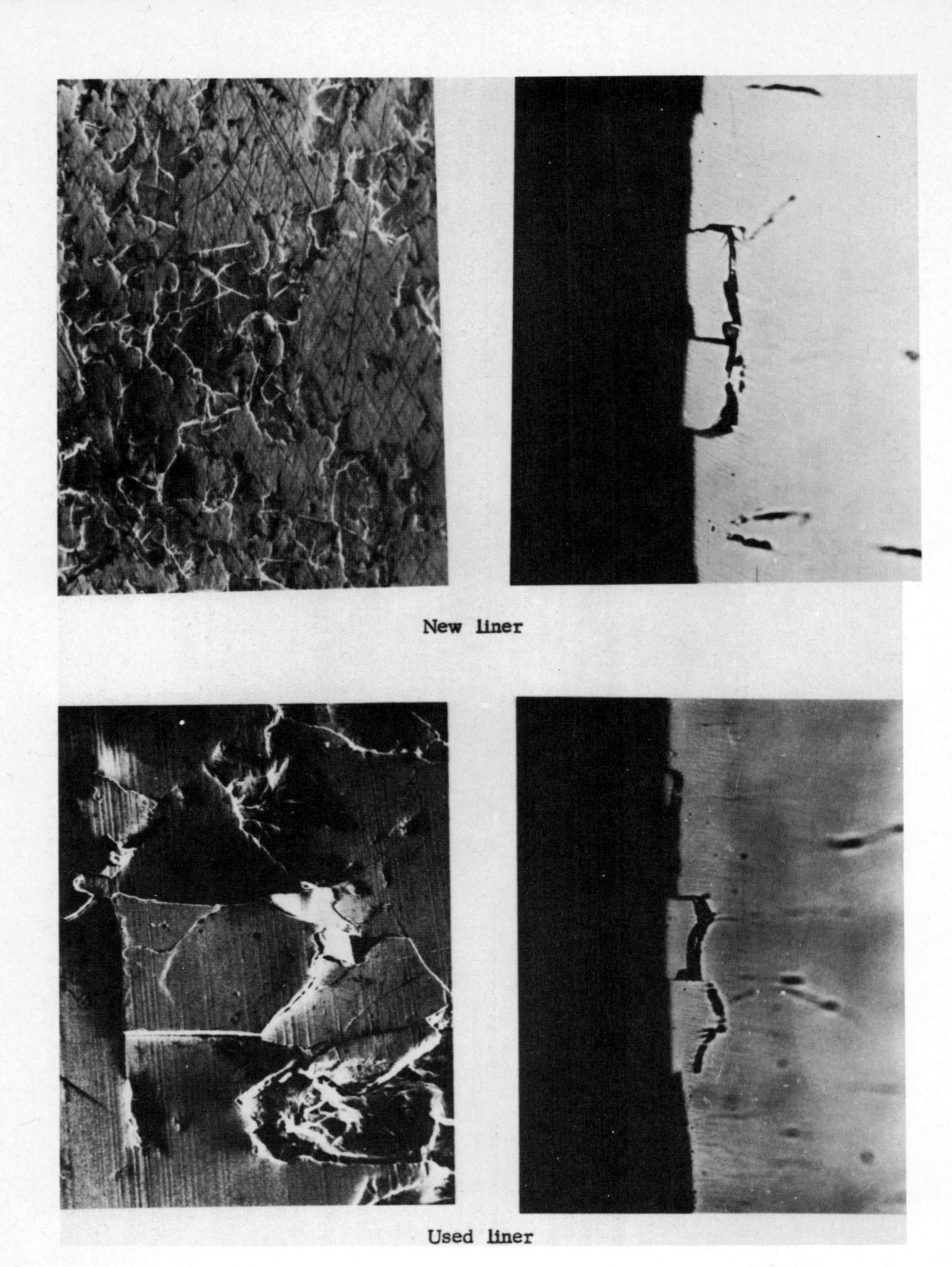

Fig 12 New and used cylinder liner surfaces showing loose particles

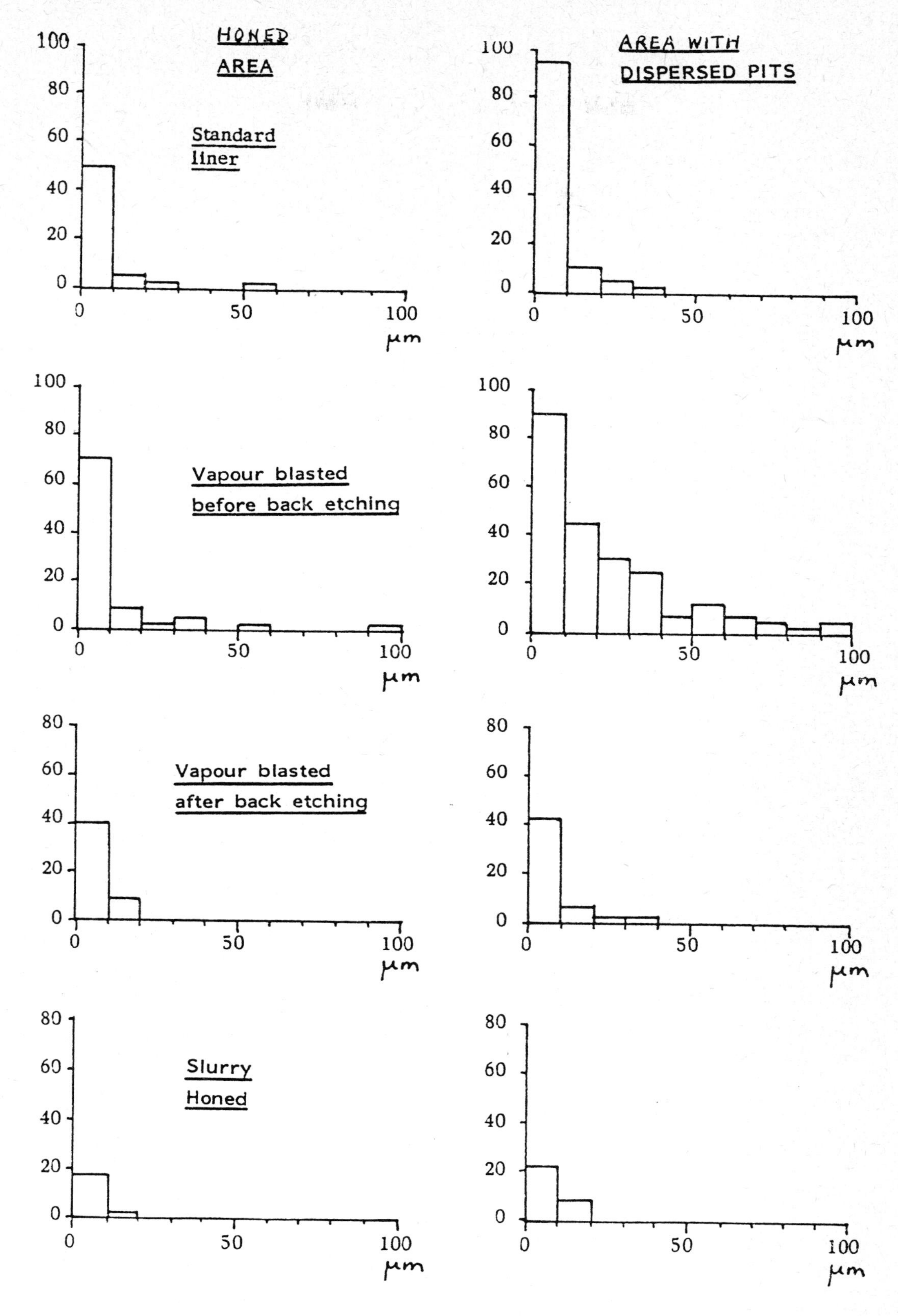

Fig 13 The number of loose particles on a 1 cm length of liner surface after various process treatments

The measurement of the boundary lubricating properties of aviation turbine fuels

J W HADLEY, BEng, CEng, MIMechE
Shell Research Limited, Thornton Research Centre, Chester

SYNOPSIS The principal types of lubrication failure occurring in high-pressure fuel pumps fitted to aircraft gas turbine engines are described. Possible preventative measures are discussed in relation to the areas of fuel usage requiring measurements of fuel boundary lubricating properties or lubricity. Candidate mechanical lubricity evaluators are listed and the three principal methods are described in detail. Comparison of friction and wear measurements in the Thornton Aviation Fuel Lubricity Evaluator for a wide range of fuel types indicate that lubrication breakdown is better indicated by a friction based failure load rather than wear scar measurement.

1 INTRODUCTION

The boundary lubricating properties or lubricity of aviation turbine fuels have in recent years come into focus in relation to potential lubrication problems in the fuel systems of aircraft gas turbines. Aviation turbine kerosine (AVTUR, Jet A, Jet A-1) is the principal fuel used by the civil market. In addition, a wide-cut fuel spanning the gasoline and kerosine boiling range (AVTAG, JP-4, Jet B) is used on a large scale by many national air forces and to a relatively small degree by the civil market. A third fuel, having a high flash point (AVCAT, JP-5), is used for naval aircraft.

Aviation turbine fuels are potentially poor lubricants by virtue of limitations placed on their composition by specifications and fuel systems rarely have an external means of lubrication. Thus, in the earliest days of the aircraft gas turbine engine, up to 3%v of lubricating oil used to be added to the fuel to provide lubrication of piston pumps.(1),(2) Such a practice is no longer permitted so that any rubbing contacts are potential sources of lubrication breakdown with the consequent deterioration in fuel system performance.

The critical fuel system components are valves and pumps. Spool valves can be sensitive to fuel lubricity by virtue of the large surface areas and small clearances involved. The sticking of such valves was experienced in some USAF aircraft in 1965. The phenomenon was caused by clay treating the fuel to remove pipeline corrosion inhibitor in order to improve water separation properties; and solved by the re-introduction of corrosion inhibitor to the fuel. The problem was investigated through the Co-ordinating Research Council (CRC) in the USA and led to the development of a spool valve simulator for the evaluation of fuel lubricity.(3)

The majority of lubrication problems since that time have occurred in high pressure fuel pumps. Those used in aircraft gas turbine engines are mainly of the piston and gear types with the latter now predominating in number. Piston pumps suffered lubrication breakdown, particularly in the late 1960s and early 1970s and on occasions since then.(4),(5),(6) The failures have taken the form of loss in pumping performance with occasional piston/bore seizures (Fig 5), leading to a loss in engine thrust. In gear pumps a number of lubrication malfunctions have been observed, the principal ones being scuffing and wear of gear teeth (Fig 6) and bearing failures. In addition to such occurrences associated with lack of lubrication, cavitation erosion of teeth and bearings and spline wear have been common causes of shortened pump lives.(7) Unlike failures in piston pumps, gear pump failures are rarely total, the usual consequences being a reduction in pumping efficiency and premature renewal.

The responses of the industry to fuel system lubrication problems were both individual, the relevant manufacturers seeking to solve their own particular problems, and co-operative through the establishment of groups of representatives from the aircraft industry. In the UK, a Lubricity Panel was set up under the supervision of the Procurement Executive of the Ministry of Defence, the controlling authority for the UK aviation fuel specifications. In the USA, the CRC have administered similar panels. These groups have sought to understand and find solutions to lubrication problems by carrying out co-operative programmes of investigation.

2 REMEDIES FOR FUEL SYSTEM LUBRICATION PROBLEMS

It should be stressed that with very few isolated exceptions the vast majority of aircraft fuel systems perform satisfactorily with the wide range of aviation fuels currently available. However, hydrocarbons vary in their lubricity and fuel systems vary in their sensitivity to fuel lubricity. At large airports, the potential problems of a fuel-sensitive pump operating on low-lubricity fuels do not generally arise because of the use of co-mingled fuel storage fed by different suppliers, with beneficial consequences readily demonstrated in the laboratory.(8) The corollary of such a situation is that the combination of a pump requiring high-lubricity fuel with a single supply of fuel having low lubricity could lead to fuel pump problems caused by poor lubrication.(4),(6) The remedies for such problems are two-fold; firstly, the sensitivity of a pump to fuel lubricity may be reduced through suitable re-design and secondly, the lubricity of fuel supplies may be increased. In practice, both pump development and changes in the fuel supply tend to be long-term solutions so that remedial action in the short term tends to be based on the use of appropriate lubricity-improving additives. Conventional extreme-pressure additives used in lubricating oil formulations would require exhaustive testing to evaluate any potential risks that compounds containing phosphorus, sulphur, chlorine or metals would introduce to other parts of the fuel and engine system.(9) Fortunately, fatty-acid type corrosion inhibitors, already approved for addition to fuels for the protection of piplelines and storage facilities, have excellent boundary lubricating properties. A number of fuel system lubrication problems have been alleviated by the addition of small concentrations of such additives to the fuel.(10) Such a procedure allows a period of grace for the industry to develop pumps that are less fuel sensitive.(11)

3 FUEL LUBRICITY MEASUREMENT

One of the objectives of the UK MOD (PE) Lubricity Panel was to define or develop a suitable lubricity measuring method. In spite of a considerable amount of work by the participating members, no such method had been identified by the time the panel was discontinued in 1976. Even now no generally accepted method of measuring fuel lubricity is available although a number of organisations are endeavouring to remedy this situation. It is therefore pertinent to consider the functions of a fuel lubricity test since it may not be possible for one method to satisfy all the possible requirements.

3.1 Measurement of the lubricity of aviation fuel supplies

The availability of a method to measure the lubricity of aviation fuel does not necessarily imply that fuel lubricity should become a specification property. However, the ability to identify those low-lubricity fuels that may be used with fuel-sensitive pumps allows the option of additive treatment of fuels to be considered. Nevertheless, it should be noted that some fuel system problems are not necessarily caused by poor fuel lubricity, for example cavitation erosion and spline wear, and may not be solved by the use of lubricity-enhancing additives.

3.2 The control of reference fuel lubricity in pump development

It is important to develop aircraft equipment to embrace the most arduous conditions likely to be found in practice. Applying this principle to fuel pump development, the fuel of the lowest lubricity presently available should be used for tests. This fact has been recognised in the UK where low-lubricity fuels are used for pump development, for example, Jet A-1, JP-4 and iso-octane,(2),(6),(11) and in the USA where iso-octane is commonly specified for aviation pump development and a mechanical lubricity evaluator is the subject of a lubricity test method for test stand fuel.(12)

3.3 The qualification of additives for use as lubricity improvers

In general, corrosion inhibitors containing long-chain fatty acids also have good boundary lubricating properties. If such materials have been approved for addition to aviation fuels for the purpose of corrosion inhibition then the requirements for their use as lubricity-enhancing materials will have largely been met. Whether such additives are required to demonstrate lubricity-enhancing properties depends on the specifying authority. In the UK, lubricity-improving materials submitted for inclusion in the fuel specification have historically been evaluated in a pump test. Currently neither the full-scale pump test nor sufficient quantities of low-lubricity reference fuel on which to operate are available. This creates the need for a suitable laboratory test method for measuring fuel lubricity. Since the fuel system temperatures can reach *ca.* 165°C it is imperative that the test should also be capable of operating at elevated temperatures.

3.4 The evaluation of alternative fuel supplies

New sources of crude and new processes for refining aviation fuels are being developed by the oil industry world-wide. Examples of new fuels are those derived from shale oil, oil sands and coal, while the thermal and catalytic cracking of heavy hydrocarbons is increasingly being utilised to increase distillate production. In view of the importance of fuel lubricity to some equipment it is necessary to evaluate all new components and compare their boundary lubricating properties with those of established fuel blends. A current example of problems that may arise from non-traditional fuel supplies is in the use by the USAF of JP-4 derived from shale. This fuel has very poor lubricity and the addition of a lubricity enhancer/corrosion inhibitor, a practice which is routine for JP-4 fuels, is essential in this case. Since depletion of the additive can occur in the fuel supply system, a mechanical lubricity evaluator is used to monitor additive concentrations in the fuel supply network.

3.5 The investigation of factors influencing fuel lubricity

Supply and operating factors such as clay treatment, which can remove boundary lubricating components from fuel; temperature, which can cause desorption of boundary lubricants from rubbing surfaces; the presence of water, which is a poor lubricant; are likely to have highly significant influences on fuel system lubrication, particularly with respect to alternative fuel supplies and lubricity-additive treatment. It is desirable therefore, to evaluate such factors in a lubricity measuring test.

4 CHOICE OF LUBRICITY TEST

The Lubricity Panel of the UK MOD (PE) conducted the most exhaustive evaluation of laboratory lubricity testing methods reported to date in attempting to identify a suitable candidate.(6) Mechanical lubricity evaluation methods were of course the major consideration and several appeared promising. Chemical methods of determining the presence of surface-active species likely to provide effective boundary lubrication were also considered but found to have serious disadvantages.

In designing a lubricity test using mechanical test rigs, it has always been the hope that a simple laboratory apparatus employing rubbing components of elementary geometry can be devised that will simulate practical machinery. Such an apparatus could significantly reduce testing costs in both fluid and component development. A multitude of simulation devices have been used and many of them have been catalogued by the American Society of Lubrication Engineers.(13) In devising or choosing a practical test device, realistic rubbing contact conditions should be simulated as far as possible. Since duplication of the practical situation can result in prolonged experimentation, tests are often accelerated by increasing the applied load; however, the mode of failure should still reflect the practical case. The accepted way of assessing the relevance of a particular test method is by comparing the performance of reference fluids in the test with their field performance, usually in terms of friction, wear, scuffing or seizure. In addition, repeatability and reproducibility should be satisfactory.

With specific reference to the field of aviation, the two most important modes of failure, already referred to, displayed by some fuel-sensitive pumps were piston/bore seizure and gear tooth scuffing and wear. The metallurgy of the former was steel sliding on cadmium-plated aluminium- bronze and that of the latter was steel rolling and sliding on steel.

The lubricity evaluators used in attempts to simulate the performance of aviation fuel pumps have included the Four Ball Machine, the Ball-on-Cylinder Machine, the Dennison Tribotester, the Lucas Dwell Test, the Esso Pin-on-Cylinder Machine and the Thornton Aviation Fuel Lubricity Evaluator. For the sake of completeness, mention should also be made of the Bendix Spool Valve Lubricity Tester developed through the CRC for investigating spool valve lubrication problems. In the case of the pump simulators, the rubbing speeds are fairly similar but the metallurgies, contact geometries and contact stresses employed in the various test methods show considerable differences (Figure 1 and Table 1).

The use of reference fluids is an essential part of any lubrication study. In the case of aviation fuels, fluids of similar viscosities and hydrocarbon types should be used and preferably they should be real fuels. Examples are solvents boiling in the kerosine range, low and high lubricity Jet A-1, JP-4 and JP-5. The lubricity may be known from field practice, inferred from previous experience of fuels from the same source, clay treated to ensure low lubricity or doped with additive to produce high lubricity.

In the present paper, discussion will be confined to the three evaluators currently in use in the aviation industry, i.e. the Ball-on-Cylinder Machine, the Lucas Dwell Test and the Thornton Aviation Fuel Lubricity Evaluator, although it should be mentioned that the Dennison Tribotester, athough not discussed here in detail has been used with some success for investigating the boundary lubricating mechanism of lubricity-enhancing additives.(14),(15)

5 DESCRIPTION OF EVALUATORS

5.1 Ball-on-Cylinder Machine

In the original form of this test used for investigating aviation turbine fuel lubricity, the wear scar diameter, friction and electrical resistance between ball and cylinder were all measured.(1),(16) Contact loads used were normally 0.24 to 4.0 kg and, in addition, the atmosphere could be varied (in some models the temperature can also be controlled). The balls were 12.7 mm diameter ball bearings as used in the Four Ball Machine, made from AISI 52100 steel, hardened to 63 Rockwell C. The specimen configuration is given in Table 1 and illustrated in Figure 1.

The cylinders (44.5 mm diameter) were made of the same material but were considerably softer, having a hardness of around 22 Rockwell C. A number of fuels, lubricants and additives were tested under a variety of environmental conditions and the results compared with those from Four Ball tests, vane pump tests and Ryder gear rig tests. A rough correlation was obtained between vane pump and gear rig tests and Ball-on-Cylinder results for a limited number of fluids. At low loads better discrimination between high- and low-lubricity fuels was given by wear scar diameter than by friction. At high loads, friction values for low-lubricity fuels were extremely erratic and wear scar repeatability deteriorated. Hence friction was discarded in favour of wear scar diameter as the significant parameter. Tests in which the environment in the Ball-on-Cylinder Machine was varied showed that, for some aviation fuels, the presence of oxygen increased wear, particularly in conjunction with water vapour and a corrosive wear mechanism was postulated. This was in contrast to experience in dry rubbing and lubricant testing where the presence of oxygen generally reduced friction and wear.(17) Extreme-pressure additives, used in lubricants, were often found to be pro-wear although fatty-acid type corrosion inhibitors provided significant anti-wear activity. Polycyclic aromatics were found to have anti-wear activity in the machine and their effect in a wider range of fluids was reported separately.(18) Subsequent to the investigations of the 1960s, a considerable amount of work was carried out in US military laboratories to investigate the performance of JP-4 and JP-5 fuels. The anti-wear effect of fatty acid types of corrosion inhibitors and the pro-wear effect of some sulphur compounds were confirmed. The effect of relative humidity on wear behaviour was quantified and this ultimately led to the recommendation that, to avoid scuffing wear in a test, a value of 10% relative humidity should be used.(19) The most recent work on the Ball-on-Cylinder Machine has been organised in the USA through the CRC. With the more widespread use of the apparatus, a major concern has been with repeatability and reproducibility and a great deal of attention is being paid to the consistency of balls and cylinders. Important uses for the ball-on-cylinder machine are the monitoring of test-stand fuel lubricity(12) and the assessment of lubricity additive concentration in shale-derived JP-4.

The conclusion from these studies is that the Ball-on-Cylinder Machine can be useful for monitoring changes in low-lubricity fuel or for estimating the concentration of a specific additive in a particular fuel. In view of the high wear obtained with some sulphur compounds, its usefulness for identifying low-lubricity fuels when supplies can contain up to 0.2%w of sulphur is questionable and would need to be positively demonstrated.

5.2 Lucas Dwell Test

The specimen configuration for this test is given in Table 1, illustrated in Figure 1 and is fully described in the literature.(4),(6) Many laboratories have evaluated the test and the specimen metallurgy used for the bulk of their programmes has been an aluminium-bronze pin rubbing on a tool-steel disc. These materials were chosen to simulate those of the axial piston pump having lubrication problems. In the pump, the bronze rotor was also cadmium-plated; however, this was found difficult to simulate in practice and the bronze pin without plating was retained, the justification being that the cadmium was removed in service and the critical rubbing pair was the bronze/steel combination. The flat pin-on-disc configuration results in low stress and, whereas all the other laboratory evaluation mentioned operate with the contacts flooded, the Dwell Test is designed to operate under starvation conditions. During the test sequence, a thin film of fuel is obtained on the rotating steel disc, the bronze pin is loaded onto it, and the frictional force on the pin is measured. The number of revolutions taken by the disc for the frictional force to reach a pre-determined level is called the dwell number. Although the method is superficially similar to the techniques used by Bowden and Tabor,(20) an extensive test programme showed that:

1. Dwell number was determined primarily by fuel volatility as broadly indicated by the final boiling point of the fuel (Figure 2).
2. A low dwell number fuel could be changed to a high dwell number fuel by the addition of small concentrations of high-boiling n-alkanes - a practice unlikely to improve fuel lubricity in pumps.(6)
3. The increase in dwell number was small when lubricity additives were added to low dwell number fuel.

These results were later corroborated by other workers.(6)

4. Some laboratories found that dwell number was relatively consistent over a long period of time and was therefore thought useful in monitoring test-stand fuel used in pump development. Since the composition of fuel over the critical high-boiling range tends to remain relatively unchanged with time, the constancy of dwell number is not surprising.

In conclusion, it may be stated that the Lucas Dwell Test is unsuitable for measuring the lubricity of aviation turbine fuels and results can be misleading, particularly if used in pump development.

5.3 Thornton Aviation Fuel Lubricity Evaluator

This apparatus is normally used in a cylinder-on-cylinder mode for the evaluation of the lubricity of fuels in Hertzian steel contacts. The two specimens, described in Table 1 and illustrated in Figure 7, are similar to one another in dimensions, 50 mm diameter, 10 mm width, and are made from medium-nickel case-hardened steel. The upper specimen has a hardness of 600 DPN and is ground to a surface finish of 0.025 μm CLA whereas the lower specimen has a harness of 850 DPN and is ground to a surface finish of 0.15 μm CLA. The upper specimen is stationary and is loaded onto the lower one which is rotated at 200 rev/min (giving a sliding speed of 0.52 m/s) resulting in a line contact characteristic of a gear pump.

The fuel supply system is a particularly important feature of the evaluator. If a static flooded system is used, oxidation of the bulk fuel takes place; likewise, if the fuel is allowed to trickle from a reservoir over the specimens to waste, oxidation is again a problem. This difficulty was overcome by containing the specimens in a small chamber through which fuel is continually passed, thus providing a single-pass flooded system.(21)

The method of operation is to apply a series of increasing loads within the range 1 to 200 kg for 15 minutes per load stage, measuring friction and fuel temperature until gross seizure occurs or the load capacity of the machine is reached. The upper specimen is turned through 5° of arc at each new load stage to present a new surface to the lower cylinder, thereby ensuring the required value of Hertzian stress. At the end of the test, wear scars are inspected and measured and coefficient of friction and wear scar width are plotted against applied load. Lubrication breakdown, observed by visual observation of the wear scars, corresponds to a large transition in coefficient of friction from steady values around 0.15 to unsteady values around 0.4 or more. For low-lubricity Jet A-1 fuel, the transition to high coefficient of friction occurs at low values of applied load, typically 12 kg. The high unsteady level of friction is sustained until gross seizure or welding occurs, usually at several load stages higher than the lubrication breakdown load, typically at 50 kg. For higher lubricity fuels, the transition to gross seizure is usually direct, but at much higher loads, typically 150 kg for a high lubricity Jet A-1. For a low-lubricity fuel containing a lubricity enhancing additive, the intermediate transition is suppressed in a similar manner to that for the high-lubricity fuel but gross seizure levels are dependent upon additive concentration, for example, at low concentrations the gross seizure load may not be raised at all. Typical plots of coefficient of friction and wear scar width against load for reference fuels are given in Figure 3. In order to quantify the lubricity performance, an arbitrary level of 0.4 in coefficient of friction has been taken as a failure criterion. From the coefficient of friction vs load plot, a derived friction failure load may then be obtained. Friction failure loads have been found to correlate well with axial piston pump performance in rig tests and with the field performance of fuel in gear pumps.(21)

When wear scar width is considered, low-lubricity fuels with and without additives give good correlation with friction failure load; however, some high-lubricity fuels also give high wear, probably by a corrosive mechanism. This is illustrated in Figure 4 which plots friction-failure load against wear scar width for a number of fuels.

The behaviour of fuels in the Thornton Aviation Fuel Lubricity Evaluator may be summarised as follows:

1. A friction-failure load, defined as the load at which the coefficient of friction reaches 0.4, gives excellent discrimination between fuels. Correlation is good with both piston-pump rig tests and the expected field performance of fuels in piston and gear pumps.
2. Gross seizure load differentiates between high- and low-lubricity fuels without lubricity-enhancing additives. However, when additives are added to low-lubricity fuel, the increase in gross seizure load is not always as high as might have been expected.
3. Wear scar width can differentiate between low-sulphur, low-lubricity fuels and between those with and without lubricity- enhancing additive. However, some high- lubricity fuels also give high wear.
   The evaluator may be used to monitor aviation turbine fuel lubricity in any of the roles previously described. It is recognised that both the volume of fuel required for full load range evaluations and the length of test can be inconvenient in some applications, however test conditions can be modified to operate on smaller fuel samples although less data would then be obtained.

6 DISCUSSION

The concern to minimise fuel pump maintenance and replacements with changing composition of aviation fuels has served to maintain a continuing interest in fuel system lubrication problems. Clearly, the availability of engine components that will operate satisfactorily on low lubricity fuel is a desirable objective; however, the difficulties associated with achieving this goal against a background of increasing thermal and mechanical stresses in engines and changes in composition of the aviation turbine fuel pool have highlighted the need for reliable methods of measuring fuel lubricity.

When considering the various tests available, it should be noted that one single method may not necessarily be suitable for all needs since each area of application can have differing requirements. This is important in fuel-pump development where a rapid test is required to check that the lubricity of the recirculated reference fuel, such as iso-octane, JP-4 and Jet A-1, is being maintained at a low level. The results obtained from this control test should correlate with the fuel lubricity as defined by more realistic tests or pump evaluation. A similar requirement is obvious when monitoring lubricity-additive concentration in JP-4 derived from shale; in this case, a single fuel type is being used, the behaviour of which is predictable with respect to additive concentration. In both of these cases the Ball-on-Cylinder Machine has been usefully applied; however, caution should be exercised in widening the use of this test to the assessment of the lubricity of fuels of widely varying composition. Although the small fuel sample required is an obvious advantage, the relationship between ball wear and fuel lubricity in practice needs to be established.

When qualifying additives for use as lubricity improvers, fuel volume and rapidity of test are of minor importance. The major requirements are the generation of detailed tribological information and the simulation of service conditions, particularly in terms of contact stress and operating temperature. The ability of the Thornton evaluator to operate at elevated temperature and over the stress range found in gear-pump tooth contacts makes it very suitable for (i) the qualification of additives, (ii) the detailed assessment of new fuel composition and (iii) the investigation of those supply and operating factors affecting fuel lubricity.

7 CONCLUSIONS

The major areas where aviation fuel lubricity measurements are required have been described in the context of aircraft engine pump design and operation. Of the three principal methods available for assessing fuel lubricity, the Lucas Dwell Test is not applicable in its original form whereas the Ball-on-Cylinder machine and the Thornton-developed Aviation Fuel Lubricity Evaluator have crucial roles to play in successful fuel systems and fuel supply development.

REFERENCES

(1) APPELDOORN, J.K. and DUKEK, W.G., Lubricity of Jet Fuels SAE Aeronautic and Space Engineering and Manufacturing Meeting, Los Angeles, California, October 3-7, 1966. SAE 660712.

(2) LOVE, B.E., HATCHETT, K.A. and PEAT, A.E., Fuel Related Problems in Engine Fuel Systems, SAE 660712.

(3) Co-ordinating Research Council, 219, Perimeter Center Parkway, Atlanta, Georgia 30346.

(4) AIRD, R.T. and FORGHAM, S.L., The Lubricating Quality of Aviation Fuels, Wear, 18 (1971) 316-380.

(5) OGLE, J., Interim Report of the Fuel Lubricity Panel 1971, Procurement Executive Ministry of Defence, AX/395/014.

(6) ASKWITH, T.C., HARDY, P.J. and VERE, R.A., Lubricity of Aviation Turbine Fuels. Second report of the work and findings of the MOD (PE) Fuel Lubricity Panel, Ref: AX/395/014, January, 1976.

(7) VALTIERRA, M.L., PAKRIS, A. and KU, P.M., Spline Wear in Jet Fuel Environment, Lubrication Engineering, 31 (March 1975) 136-142.

(8) VERE, R.A., Dilution Restores Lubricity, SAE Journal, 78 (4) (1978) 42-3.

(9) SMITH, M., Aviation Fuels, G.T. Foulis & Co. Ltd.

(10) VERE, R.A., Lubricity of Aviation Turbine Fuels, SAE 690667.

(11) HAMILTON, L.P. and SPARKS, B.E., Pumps for Low Lubricity and Corrosive Fuels, ASME Gas Turbine Conference, Houston, Texas, March 2-6, 1975.

(12) Aircraft recommended practice ball-on-cylinder (BOC) aircraft turbine fuel lubricity tester, SEA MAP 1794.

(13) Friction and Wear Devices, ASLE 2nd Edition, 1976.

(14) POOLE, W. and SULLIVAN, J.L., The wear of aluminium-bronze on steel in the presence of aviation fuel, ASLE Trans, 22 (2) (1979) 154-161.

(15) POOLE, W. and SULLIVAN, J.L., The role of aluminium segregation in the wear of aluminium-bronze/steel interfaces under conditions of boundary lubrication, ASLE Trans., 23 (4) 1980 401-408.

(16) TAO, F.F. and APPELDOORN, J.K., The ball on cylinder test for evaluating jet fuel lubricity, ASLE Trans., 11 (1968) 345-352.

(17) VINOGRADOV, G., KOREPOVA, I.V., PODOLSKY, Y.Y. and PAVOLOVSKAYA, N.T., Effect of Oxidation on Boundary Friction of Steel in Hydrocarbon Media and Critical Friction Duties Under Which Cold and Hot Seizure (or Welding) Develop, ASME/ASLE International Lubrication Conference, Washington, DC., October 13-16, 1964.

(18) APPELDOORN, J.K. and TAO, F.F., The Lubricity Characteristics of Heavy Aromatics, Wear, 12 (1968) 117-130.

(19) GRABEL, L., Lubricity properties of high temperature jet fuel, Report NAPTC-PE-112. Naval Air Propulsion Test Centre, Trenton, New Jersey, August, 1977.

(20) BOWDEN, F. and TABOR, D., The Friction and Lubrication of Solids, Oxford, 1950.

(21) HADLEY, J.W., A method for the evaluation of the boundary lubrication of aviation turbine fuels, Wear, To be published.

Table 1 Summary of operating parameters for mechanical lubricity evaluators

Machine	Lubrication system	Metallurgy	Component geometry	Mean contact stress, MPa	Rubbing speed, ms^{-1}	Operating temperature, °C	Reference
Ball-on-Cylinder	50 ml static	ball-AISI 52100 Rc 63 cylinder-AISI 52100 Rc 20-22	ball-12.7 mm dia. cylinder-44.5 mm dia.	565 at 1 kg	0.5	Room temperature	16
Lucas Dwell Test	1 ml centrifuged off	pin-DTD 197A aluminium bronze disc-KE961 750 DPN	pin-1 mm dia. disc-102 mm dia.	0.82	0.4	Room temperature	4,6
Thornton Aviation Fuel Lubricity Evaluator	Single pass flooded 1.1 lh^{-1}	upper cylinder - En33, 600 DPN lower cylinder - En33, 850 DPN	cylinders - 50 mm dia. 10 mm wide	43 to 612	0.52	Room temperature to 150°C	21
Four Ball	8 ml static	balls-AISI 52100 Rc 62-64	upper ball-12.7 mm dia. three lower balls-12.7 mm dia.	1800 at 10 kg	0.58	Room temperature	5,13
Dennison Tribotester	Trickle feed recirculated 1.8 lh^{-1}	pin - DTD 197A aluminium bronze disc-KE961 steel 750 DPN	pin - 120° Cone, up to 3 mm dia. disc-100 mm dia.	1 to 14	1.5	Room temperature	6,14
ESSO Pin-on-Cylinder	50 ml static	pin-silver plated steel cylinder-S15 steel	pin-6.35 × 6.35 mm cylinder-44.5 mm dia.	0.21	1.16	Room temperature	6,10

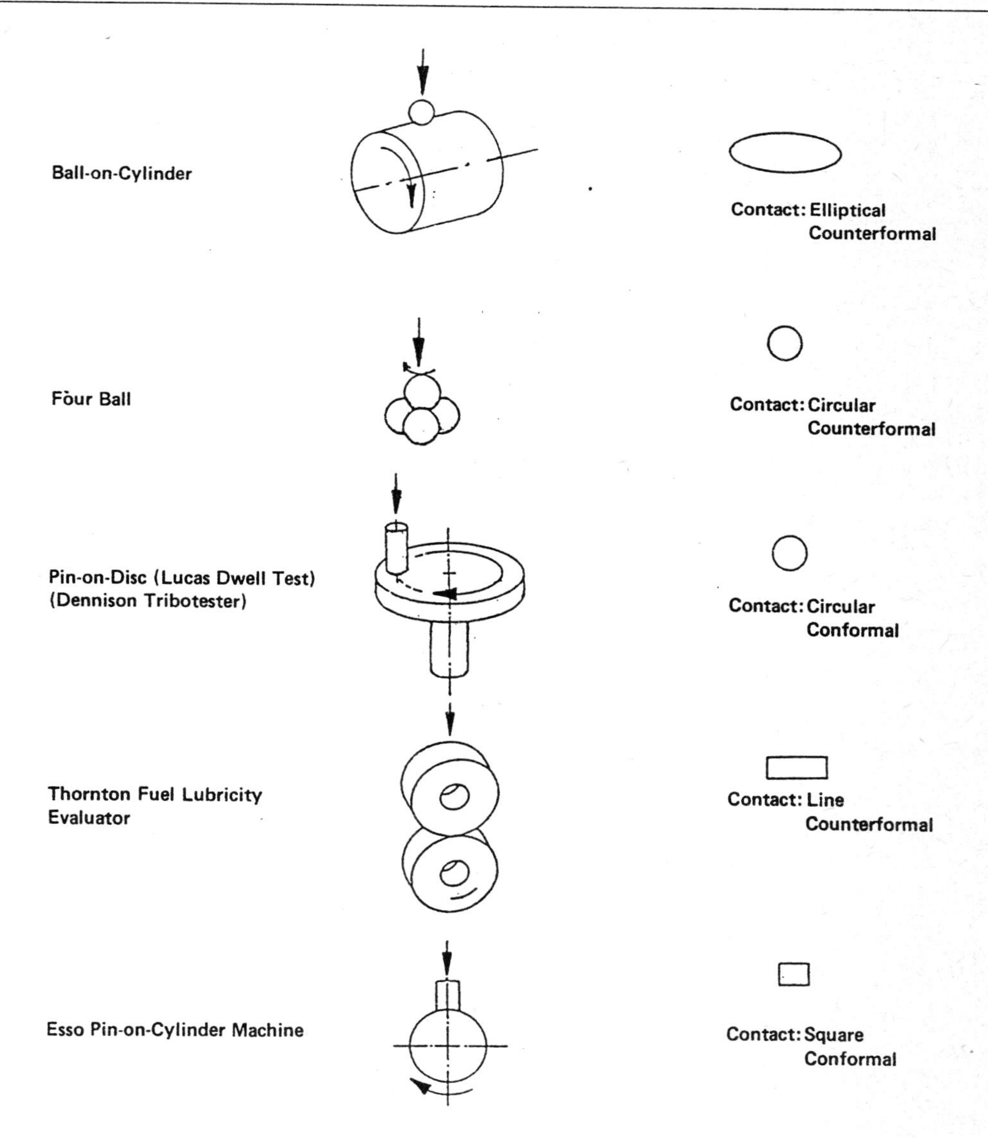

Fig 1 Contact geometry for mechanical lubricity evaluators

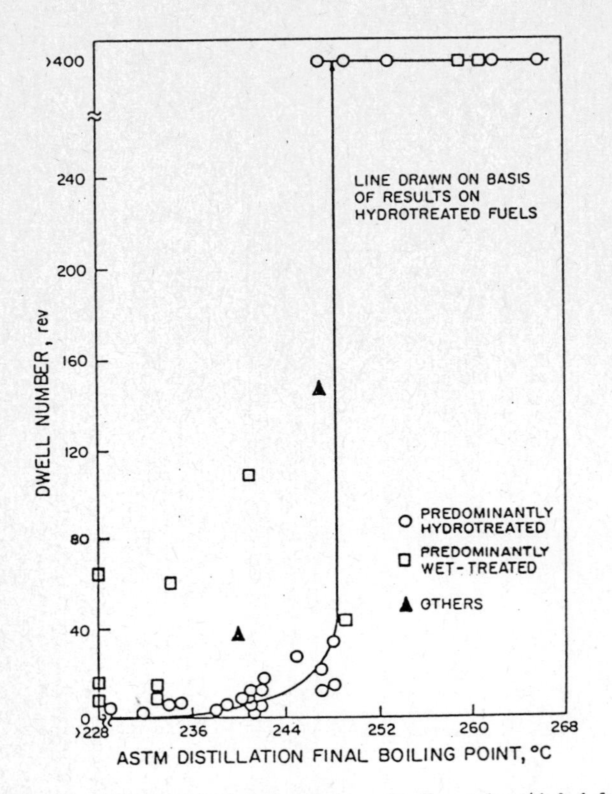

Fig 2 The effect of fuel end-point on dwell number (J A-1 fuel)

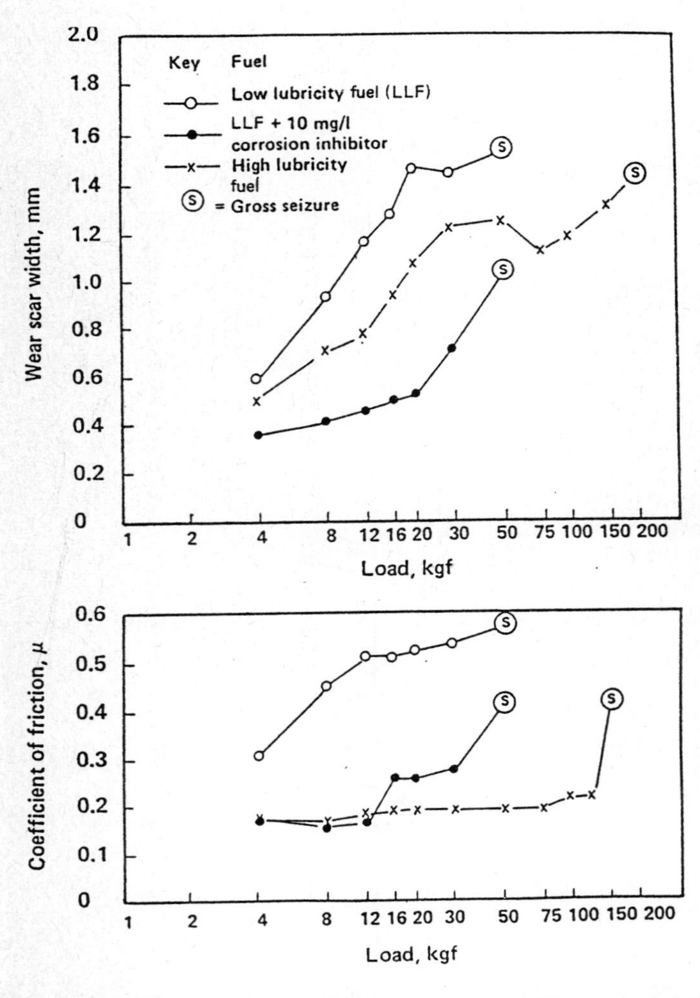

Fig 3 Behaviour of typical reference fuels in the Thornton Aviation Fuel Lubricity Evaluator

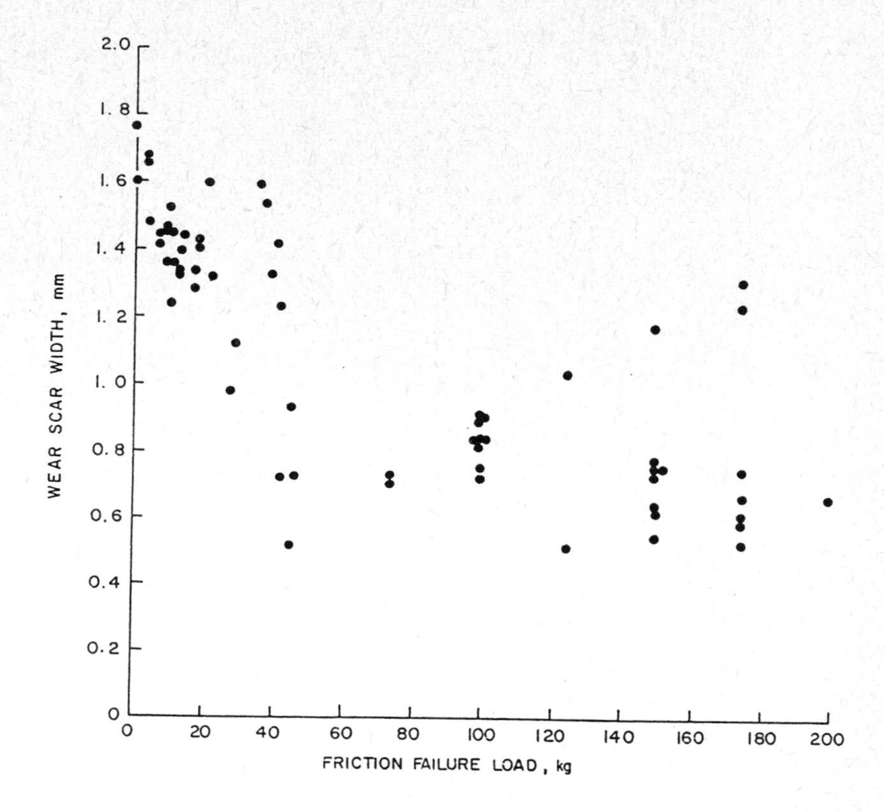

Fig 4 Graph of friction failure load versus wear scar width for different fuels in the Thornton Aviation Fuel Lubricity Evaluator

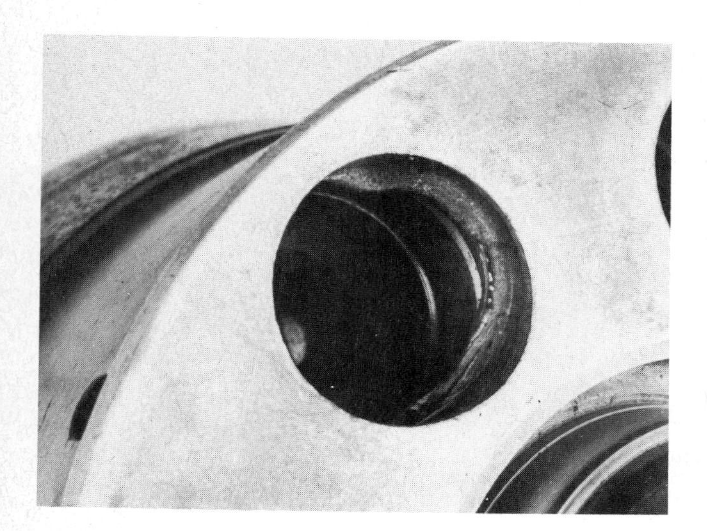

Fig 5 Piston pump lubricity failure

Fig 6 Gear pump scuffing failure

Fig 7 Test section of the Thornton Aviation Fuel Lubricity Evaluator